MODERN BATTERY ENGINEERING
A Comprehensive Introduction

MODERN BATTERY ENGINEERING
A Comprehensive Introduction

Editor

Kai Peter Birke
University of Stuttgart, Germany

NEW JERSEY · LONDON · SINGAPORE · BEIJING · SHANGHAI · HONG KONG · TAIPEI · CHENNAI · TOKYO

Published by

World Scientific Publishing Co. Pte. Ltd.
5 Toh Tuck Link, Singapore 596224
USA office: 27 Warren Street, Suite 401-402, Hackensack, NJ 07601
UK office: 57 Shelton Street, Covent Garden, London WC2H 9HE

Library of Congress Cataloging-in-Publication Data
Names: Birke, Kai Peter, editor.
Title: Modern battery engineering : a comprehensive introduction / edited by Kai Peter Birke,
  University of Stuttgart, Germany.
Description: New Jersey : World Scientific, [2019] | Includes bibliographical references and index.
Identifiers: LCCN 2018061454 | ISBN 9789813272156 (hardback : alk. paper)
Subjects: LCSH: Electric batteries.
Classification: LCC TK2896 .M63 2019 | DDC 621.31/2424--dc23
LC record available at https://lccn.loc.gov/2018061454

British Library Cataloguing-in-Publication Data
A catalogue record for this book is available from the British Library.

First published 2019 (Hardcover)
Reprinted 2019 (in paperback edition)
ISBN 978-981-121-598-8 (pbk)

Copyright © 2019 by World Scientific Publishing Co. Pte. Ltd.

All rights reserved. This book, or parts thereof, may not be reproduced in any form or by any means, electronic or mechanical, including photocopying, recording or any information storage and retrieval system now known or to be invented, without written permission from the publisher.

For photocopying of material in this volume, please pay a copying fee through the Copyright Clearance Center, Inc., 222 Rosewood Drive, Danvers, MA 01923, USA. In this case permission to photocopy is not required from the publisher.

For any available supplementary material, please visit
https://www.worldscientific.com/worldscibooks/10.1142/11039#t=suppl

Desk Editors: Gregory Lee/Amanda Yun

Typeset by Stallion Press
Email: enquiries@stallionpress.com

Contents

Preface   xiii

About the Editor   xv

About the Authors   xvii

1. **Fundamental Aspects of Achievable Energy Densities in Electrochemical Cells**   1

   Kai Peter Birke and Desirée Nadine Schweitzer

   Annex 19
   A. Specific capacity of each element 19
   B. Series voltage of each element 22
   C. Specific energy of each element 24
   D. Volumetric energy density of each element 27
   Bibliography 30

2. **Lithium-ion Cells: Discussion of Different Cell Housings**   31

   Kai Peter Birke and Shkendije Demolli

   2.1 Cell Housings 31
   2.2 Cylindrical Cells 32
   2.3 Prismatic Cells 32
   2.4 Stabilization of Electrode and Separator Layers 35
   2.5 Gas Evolution 37
   2.6 Flexibility with Respect to Cell Size 38

2.7	Producing Pouch Cells		38
2.8	Status Quo of Cell Concepts		39
2.9	Outlook		40
Bibliography			41

3. Integral Battery Architecture with Cylindrical Cells as Structural Elements 43

Christoph Bolsinger, Marcel Berner and Kai Peter Birke

3.1	State of the Art Battery Systems		45
	3.1.1	Block architecture	45
	3.1.2	Modular architecture	46
	3.1.3	Cell circuitry	46
3.2	The Battery Cell as a Structural Element		47
	3.2.1	Cylindrical cells	48
	3.2.2	Prismatic cells	49
	3.2.3	Battery cells as structural elements	49
3.3	Construction of the Battery Module		51
	3.3.1	Cell connection	51
	3.3.2	Moisture proof	52
	3.3.3	Lifetime	52
	3.3.4	Automotive standards	52
	3.3.5	No further load bearing elements	53
	3.3.6	Thermal management	54
	3.3.7	Safety aspects	54
	3.3.8	Scalability	55
	3.3.9	Exchangeable single battery cells	55
	3.3.10	Gas channels	56
3.4	Integrated Cell Supervision Circuit		56
	3.4.1	Balancing	57
	3.4.2	Mechanical integration	58
	3.4.3	Communication	58
	3.4.4	Energy saving	59
3.5	Cell Connectors		60
	3.5.1	State of the art	60
	3.5.2	Electrical contact resistance	61
	3.5.3	Clamped cell connectors	63
	3.5.4	Conclusion	65
3.6	Battery Thermal Management		66

	3.6.1	State of the art	67
		3.6.1.1 Air cooling for BTM	67
		3.6.1.2 Liquid cooling for BTM	69
		3.6.1.3 Phase change materials for BTM	70
		3.6.1.4 Heat pipe	71
		3.6.1.5 Thermoelectric cooler (TEC)	72
	3.6.2	BTM for integral single cell	74
		3.6.2.1 Non-uniform temperature distribution inside battery cells	74
		3.6.2.2 Terminal cooling	75
Acknowledgment			77
Bibliography			77

4. Parallel Connection of Lithium-ion Cells — Purpose, Tasks and Challenges   81

Alexander Fill

4.1	Introduction	81
4.2	Main Issues and Challenges	82
4.3	Influences on the Current Distribution	83
	4.3.1 Simplified model — Analytical solution	84
	4.3.2 Effects of cell resistance and capacity variations	90
	4.3.3 Influence of the open circuit voltage bending	94
4.4	Thermal Effects	97
4.5	Aging	98
Bibliography		100

5. Fundamental Aspects of Reconfigurable Batteries: Efficiency Enhancement and Lifetime Extension   101

Nejmeddine Bouchhima, Matthias Gossen and Kai Peter Birke

5.1	Introduction	101
5.2	Modeling	103
	5.2.1 Energy efficiency	103
	5.2.1.1 Energy loss	104
	5.2.1.2 Rest energy versus equalization energy	104
5.3	Dynamic Optimization Problem	105

	5.4	Optimal Control 107
		5.4.1 Vector-based dynamic programming 107
		5.4.2 Complexity of the control strategy 108
		5.4.3 Optimal control policy 110
	5.5	Efficiency Enhancement 110
		5.5.1 Simulation setup 111
		5.5.2 Results 112
	5.6	Lifetime Enhancement 114
		5.6.1 Aging model 115
		5.6.2 Results 115
	5.7	Summary 117
	Bibliography 118

6. Volume Strain in Lithium Batteries   121

Jan Patrick Singer and Kai Peter Birke

	6.1	Introduction 121
	6.2	Fundamentals of Volume Strain 121
		6.2.1 Intercalation 123
		6.2.2 Alloying 124
		6.2.3 Conversion 125
	6.3	Volume Strain on Cells Level 125
	6.4	Volume Strain on Systems Level 126
	6.5	Measurement Techniques 128
		6.5.1 Unpressurized 130
		6.5.2 Pressurized 133
	6.6	State Diagnostics 135
		6.6.1 SoH diagnostics 135
		6.6.2 SoC diagnostics 136
	Bibliography 138

7. Every Day a New Battery: Aging Dependence of Internal States in Lithium-ion Cells   141

Severin Hahn and Kai Peter Birke

	7.1	Operation and Degradation Processes in the Electrode State Diagram 141
		7.1.1 Introduction 141
		7.1.2 Absolute potentials and the electrode state diagram 142

		7.1.3	Charge and discharge	144

		7.1.3	Charge and discharge	144
		7.1.4	Charge and discharge limits	145
		7.1.5	Combined electrode reactions	146
		7.1.6	Anodic side reactions — Growth of solid electrolyte interface (SEI)	148
		7.1.7	Cathodic side reactions — Possible formation of solid permeable interface (SPI)	151
		7.1.8	Transition metal dissolution	152
		7.1.9	Loss of active material	154
	7.2	Experimental Verification and Analysis Techniques		155
		7.2.1	Loss of anode active material	156
		7.2.2	Loss of active lithium	157
		7.2.3	Loss of cathode active lithium	158
		7.2.4	The principle of limitation	158
		7.2.5	Example of an aged cell	159
		7.2.6	Inhomogeneities and limitations in real cells	160
	7.3	Conclusion		161
	Bibliography			163

8. Thermal Propagation   167

Sascha Koch

8.1	Introduction		167
8.2	Process of Thermal Propagation		167
	8.2.1	Thermal runaway	167
	8.2.2	Propagation	169
	8.2.3	Resulting effects	171
8.3	Testing		172
	8.3.1	Relevance	172
	8.3.2	Trigger methods	172
	8.3.3	Measurement equipment and methods	174
	8.3.4	Experiment setup and conditions	177
	8.3.5	Analyzing the results	178
8.4	Influencing Variables		181
	8.4.1	Cell format	181
	8.4.2	Energy density	182
	8.4.3	System design	183
Bibliography			184

9. Potential of Capacitive Effects in Lithium-ion Cells — 187
Alexander Uwe Schmid and Kai Peter Birke

- 9.1 Brief Introduction to the Principles of Electrostatic and Electrochemical Storage 187
  - 9.1.1 Double-layer capacitance 188
  - 9.1.2 Intercalation 190
  - 9.1.3 Pseudocapacitance 190
- 9.2 Similarities and Differences between Capacitors and Lithium-ion Cells 191
  - 9.2.1 Carbons as electrode material 192
  - 9.2.2 The solid electrolyte interface 193
  - 9.2.3 Summary 194
- 9.3 Methods of Measurement of Capacitive Effects 195
  - 9.3.1 Electrochemical impedance spectroscopy 195
    - 9.3.1.1 Modeling approaches based on equivalent circuit elements 196
  - 9.3.2 Cyclic voltammetry 203
  - 9.3.3 Current pulse method 204
  - 9.3.4 Summary 205
- 9.4 Utilization of Capacitive Effects in Li-ion Cells 205
  - 9.4.1 Li-ion cell development 205
  - 9.4.2 Li-ion capacitor 206
  - 9.4.3 Estimation of DL capacitance on cell level 207
  - 9.4.4 Potential on the system level 211
- 9.5 Conclusion and Outlook 216
- Nomenclature 217
- Bibliography 220

10. Battery Recycling: Focus on Li-ion Batteries — 223
Daniel Horn, Jörg Zimmermann, Andrea Gassmann, Rudolf Stauber and Oliver Gutfleisch

- 10.1 Battery Materials and their Supply 223
- 10.2 Motivation for Battery Recycling and Legal Framework in Europe 227
- 10.3 Available Recycling Technologies 228
  - 10.3.1 Pre-processing treatments 229
  - 10.3.2 Pyro- and hydrometallurgy for extraction 231

10.4	Electrohydraulic Fragmentation, an Innovative Recycling Process for Battery Recycling	234
10.5	Outlook	236
	Bibliography	236

11. Power-to-X Conversion Technologies — 239

Friedrich-Wilhelm Speckmann and Kai Peter Birke

11.1	Definition of Power-to-X	239
11.2	Potential of Cross-Sectoral Applications	239
11.3	Power-to-X as a Primary Battery	244
11.4	Power-to-Gas	244
	11.4.1 Hydrogen generation	245
	11.4.2 Electrolytic hydrogen generation	245
	11.4.2.1 Thermochemical hydrogen generation	248
	11.4.2.2 Photochemical hydrogen generation	249
	11.4.3 Methanation	250
	11.4.3.1 Catalytic/chemical methanation	250
	11.4.3.2 Biological methanation	251
	11.4.3.3 Plasma-based methanation	251
11.5	Power-to-Liquid	252
	11.5.1 Technological overview	252
	11.5.2 Carbon sources	255
11.6	Power-to-Solid	256
11.7	Basic Gas Management Systems	258
11.8	Sustainable Energy Chains — Closing Remarks	259
	Bibliography	260

Epilogue — 263

Acknowledgments — 275

Index — 277

Preface

A famous and well-known company created the great slogan, "Research makes the difference". To apply this in the successful development of modern batteries, it should be slightly paraphrased as "Engineering makes the difference", which in one sentence summarizes the magic formula for batteries.

This richly illustrated book written by Professor Kai Peter Birke and several co-authors embodies this by combining both scientific and engineering aspects of modern batteries in a unique approach. Emphasizing the engineering part of batteries, the book acts as a compass towards next generation batteries for automotive and stationary applications. It provides answers to still open questions on how future batteries may look like and where the journey goes for them.

The book is thus highly important in the field and market of electrical energy storage systems, commonly known as modern batteries, since it targets why and how such batteries have to be designed for successful commercialization in e-mobility and stationary applications. It helps readers to understand the principle issues of battery designs, paving the way for engineers to avoid wrong paths and settle on appropriate cell technologies for next generation batteries. In the same time, this book is ideal for advanced training courses for readers interested in the field of modern batteries.

Modern Battery Engineering provides a unique treatment of selected aspects of battery engineering. It highlights new material and the state-of-the-art in the field as a result of the joint comprehensive knowledge of active researchers and engineers in the field of batteries.

Stuttgart, August 14th, 2018
Kai Peter Birke

About the Editor

Professor Kai Peter Birke is a physicist and a full Professor at the University of Stuttgart, Germany, covering the field of Electrical Energy Storage Systems, including new cell materials and technologies, advanced Li-ion batteries and Power-to-X. He obtained his Ph.D. in Materials Science (ion conducting ceramics) from the University of Kiel, Germany, in 1998. In 1999, he joined the Fraunhofer-Institute for Silicon Technology, Itzehoe, Germany, to work on the development of proprietary Li-ion laminated cells with a novel functional ceramic separator. He also co-founded two spin-offs to put this technology into production.

After being involved in the development and production of pouch type laminated (PoLiFlex) Li-ion cells in leading positions at VARTA for five years, Prof. Birke joined Continental AG, Business Unit Hybrid Electric Vehicle, in Berlin in 2005, as a Project Leader in Energy-storage Systems. He subsequently became Senior Technical Expert in Battery Technology and Team Leader in Cell Technology, and in 2010 he was appointed Head of Battery Modules and Electromechanics. In 2013, he joined the JV SK-Continental e-motion as Head of Advanced Development.

Professor Birke has 24 years of experience in research, development and production of energy storage systems with a particular focus on Li-ion technology.

About the Authors

Marcel Berner is Director of Development at Innovative Pyrotechnik GmbH in Ehningen, Germany. He received his Diploma in Electrical Engineering from the University of Stuttgart, Germany, in 2012. There, he joined the Sensor Technologies group under the supervision of Prof. Dr. Jürgen H. Werner at the Institute for Photovoltaics. His work as a Ph.D. student focuses on the fabrication of thin film photodetectors and their application in medical drug monitoring systems. Marcel has also worked on high-power modulation principles of photovoltaic modules, differential brightness measurements based on optical modulation and on optical communication in bus-controlled battery systems.

Christoph Bolsinger received his Master's degree in Electrical Engineering from the University of Stuttgart in 2015. In the same year, he joined the Chair for Electrical Energy Storage Systems under the supervision of Prof. Dr. Kai Peter Birke at the Institute for Photovoltaics, where he is currently also pursuing his Ph.D. His current research interests include the structural integration of lithium-ion cells in a modern battery system, with a special focus on their electrical, mechanical and thermal interfaces.

Nejmeddine Bouchhima is currently a System Engineer for Battery Storage Systems. He received his Diploma in Mechanical Engineering from Karlsruhe Institute of Technology (KIT) in 2012. He is persuing his Ph.D. under the supervision of Prof. Dr. Birke, which includes research work as part of a cooperation between the Chair for Electrical Energy Storage Systems, University of Stuttgart, and Daimler AG. His research interests include the development of energy equalization strategies to optimize the energy efficiency and lifetime of multi-cell batteries. His technical areas are in establishing system specification and the design of architecture and functions of high voltage batteries for electric vehicles.

Alexander Fill studied renewable energy at the Friedrich-Alexander University, Erlangen-Nürnberg, Germany, and received his Master's degree in 2017. From October 2013 to April 2017, he worked in the field of fuel cell development at Daimler AG in Untertürkheim and gained expertise in electrochemistry. Since May 2017, he has been working on his Ph.D. under the supervision of Prof. Dr. Birke at Daimler AG. His research investigates the arising effects of parallel-connected lithium-ion cells.

Andrea Gassmann is Head of the Strategy and Networks Division at the Fraunhofer Recycling and Resource Strategies IWKS in Hanau and Alzenau, Germany. There, she is engaged in the development of sustainable resource management concepts and in the assessment of environmental impacts. Previously, Dr. Gassmann developed, among other things, recycling concepts for LED lighting. Prior to this, she studied Materials Science and finished her Ph.D on the topic of cathode development for organic light-emitting diodes at the Technical University of Darmstadt (TU Darmstadt), Germany. Between 2010 and March 2015 she worked as a post doctorate and had a work group on Organic Electronics at TU Darmstadt.

About the Authors

Oliver Gutfleisch is a Professor for Functional Materials at TU Darmstadt and a scientific manager at Fraunhofer IWKS Materials Recycling and Resource Strategies. His scientific interests span new materials for power applications to permanent magnets for e-mobility, to solid state energy efficient magnetic cooling and ferromagnetic shape memory alloys with a particular emphasis on tailoring structural and chemical properties on the nanoscale. His work focuses are on resource efficiency on element, process and product levels, as well as the recycling of strategic metals. He has published more than 360 papers in refereed journals and was awarded an ERC Advanced Grant (Cool Innov) in 2017.

Matthias Gossen received his B.Sc. in Electrical Engineering, Information Technology and Computer Engineering from RWTH Aachen University, Germany, in 2012. In 2015, he received his M.Sc. from the same institute, this time focusing on power electronics, electrical drives and energy storage systems. He is currently pursuing his Ph.D. degree in Electrical Engineering at RWTH Aachen while working as a doctoral student at Deutsche ACCUMOTIVE GmbH, a company of Daimler AG. His research interests include energy storages, lithium-ion batteries and system aging.

Daniel Horn studied "Advanced Materials" at Justus-Liebig-University Giessen, Germany, where he attained his M.Sc. degree in 2014 with a process evaluation on crack formation in metal alloys. After graduating, he worked as a Research Associate at the Fraunhofer Project Group Materials Recycling and Resource Strategies. Since 2017, Daniel leads and works on projects in the department of Energy Materials and Lightweight Technology, especially in the field of battery recycling.

Desirée Nadine Schweitzer received her B.Eng. degree in Electrical Engineering from Hochschule Esslingen, Germany, in 2016. She is now studying electric mobility at the University of Stuttgart. As a working student, she has been engaged in research on power electronics with the Robert Bosch GmbH from 2016 to 2017. In October 2017, she joined a research project of motion control at Daimler AG — Mercedes Benz Cars. She started her master's thesis on electromagnetic compatibility in June 2018 and will graduate in November 2018.

Shkendije Demolli is an engineer in the field of energy technology. She received her B.Sc. in Mechanical Engineering from Technical University of Braunschweig, Germany, and her M.Sc. in Energy Technology from the University of Stuttgart. During her studies, she worked on different fields, such as explosion prevention for industrial equipment, and research in fuel cells and energy efficiency. Since May 2017, Shkendije investigates in-situ hydrogen evolution with chemical reactions at the Chair for Electrical Energy Storage Systems, University of Stuttgart.

Severin Hahn is currently a Ph.D. student under the supervision of Prof. Dr. Birke at Daimler AG. The focus of his current research is the quantitative prediction of lithium-ion battery lifetime, both on cell and system level. Prior to this, Severin studied Mechanical Engineering from 2009 to 2016 at ETH in Zurich, Switzerland, during which he completed several projects regarding flame synthesis of lithium-ion battery materials such as CuO and $LiFePO_4$. He also spent 18 months at Bosch (Corporate Research) in Renningen, Germany, where he investigated the electrochemistry, cell design and microstructure hybrid supercapacitors. Severin has authored several patents regarding hybrid supercapacitors.

Sascha Koch studied Electrical Engineering and Information Technology at the University of Stuttgart from 2010 to 2016. As part of an overseas study program, he spent two semesters, from August 2014 to April 2015, at the San Jose State University, CA, US, where he gained broader knowledge in the fields of mechatronics and microcontroller programming. Majoring in Micro-, Opto- and Power Electronics, he received his M.Sc. degree in April 2016. Since then, Sascha has been working on his Ph.D. under the supervision of Prof. Dr. Birke at Daimler AG. He is currently doing research in the field of thermal propagation in lithium-ion traction batteries.

Alexander Uwe Schmid is a Scientific Assistant currently pursuing his Ph.D., consisting of investigations about the capacitive effects and aging of lithium-ion cells, under the supervision of Prof. Dr. Birke. He joined the Chair for Electrical Energy Storage Systems, University of Stuttgart, in February 2016. Prior to this, Alexander attained his M.Sc. (Hon) in Sustainable Electrical Power Supply at the University of Stuttgart, and won the Anton- and Klara-Röser prize. He also received a scholarship from Netze BW GmbH, an electrical grid operator from southern Germany.

Jan Patrick Singer is a Research Assistant in the field of Electrical Energy Storage Systems, at the Chair for Electrical Energy Storage Systems, University of Stuttgart, focusing on the detection of electrochemical effects induced by volume strain and developing new non-destructive characterization methods for lithium-ion cells in his Ph.D. work under the supervision of Prof. Dr. Birke. He started his professional career in 2006 as a mechatronic technician at Harman/Becker Automotive Systems. In 2012, he received his B.Eng. degree in Energy Systems from the University of Applied Sciences Ulm. He later graduated with a M.Sc. in Sustainable Electrical Power Supply from the University of Stuttgart in 2015.

Friedrich-Wilhelm Speckmann received his B.Eng. degree in Electrical Engineering from the University of Applied Science Bielefeld, Germany, in 2013, and his M.Sc. degree in Sustainable Energy Distribution from the University of Stuttgart in 2015. In 2015, he joined the Chair for Electrical Energy Storage Systems under the supervision of Prof. Dr. Birke at the Institute for Photovoltaics. His work as a Ph.D. student focuses on the field of large-scale energy storage. He is involved in research regarding different sources of hydrogen generation and the field of power electronics and electric micro-grids, especially, the use of electrolysis as a source of hydrogen for further methanation via a plasma-based system.

Rudolf Stauber has been Managing Director of the Materials Recycling and Resource Strategies Fraunhofer Project Group in Alzenau and Hanau since May 2012. Prior to this, Dr. Stauber studied Chemistry in Würzburg and Mainz. He obtained his Ph.D. in 1979 in the field of Organic and Analytical Chemistry. From 1979 to 2012, he worked at the BMW Group München where he was Vice President for Materials Development and Operational Stability in Automotive Engineering. His current work priorities are resources-strategic consulting of industry and politics, the development of recycling cycles and the recovery of valuable materials, and the development of functional materials for the substitution of critical materials.

Jörg Zimmermann is a Materials scientist and Head of the Energy Materials and Lightweight Technology department at the Fraunhofer Project Group IWKS. He studied in the Materials and Earth Sciences department at TU Darmstadt, Germany, and obtained his Ph.D. on x-ray storage phosphors in 2005. From 2005 to 2014, he headed the Inorganic Phosphors working group in the Electronic Materials division of the same department. In 2014, Dr. Zimmermann became manager of the business unit Lighting at the Fraunhofer Project Group Materials Recycling and Resource Strategies IWKS in Hanau and Alzenau.

Chapter 1

Fundamental Aspects of Achievable Energy Densities in Electrochemical Cells

Kai Peter Birke and Desirée Nadine Schweitzer

Currently, lithium-ion batteries are the most common energy sources for today's portable devices like laptops and smartphones, but their fields of application extend to more demanding sectors such as the automotive and smart grid sectors. The concept of electrochemically rechargeable energy storage is still quite complex, though many simplifications of cell chemistry and suppression of undesirable side reactions have already taken place after its transition from lead acid to Li-ion.

Thus, the improvement of future secondary battery systems seems, at first glance, to be down to its choice of basic cell elements, especially active materials that contain energy.

In this regard the following three main cell characteristics have to be considered: (1) the specific capacity, (2) the specific energy density and (3) the volumetric energy density. However, one important question remains: will the most promising active materials result in the most appropriate battery? In this chapter, a suitable method to collect the best basic cell active materials for a battery will be shown.

The specific capacity, typically provided as [mAh/g] or [Ahkg^{-1}] of an element, shows how much electricity can be stored in one gram of this element. This depends on the number of electrons z that can be transferred in the electrochemical redox reaction [1]. The electrical charge per mole of electrons is defined as the Faraday constant F, which gives, if multiplied with the number of electrons z, the molar capacity:

$$C^M = z \cdot F. \tag{1.1}$$

The value of the Faraday constant is about $96\,485\,\text{C}\,\text{mol}^{-1}$. This molar capacity for the given electrochemical redox reaction as a function of the molar mass of the element and the number of transferred electrons (this is the redox-reaction specific part) results in the specific capacity of the single element or chemical compound, as shown in Eq. (1.2). To calculate the specific capacity of a single element, it is necessary to take the ideal assumption that all of the material participates in the electrochemical redox reaction. The value of the molar mass of each element M_{AM} (AM = Active material) can be easily derived from the periodic table [2].

$$C^m = \frac{C^M}{M_{AM}} = \frac{z \cdot F}{M_{AM}}. \tag{1.2}$$

As 1 Coulomb = 1 Ampere × 1 second, it is recommended to divide Eq. (1.2) by the factor 3.6 to obtain the specific capacity C^m in the appropriate unit $\left[\frac{\text{mAh}}{\text{g}}\right]$.

The atomic mass is the mass of an atom, which can be expressed in kilograms [kg] (See Eq. (1.3)). Often, the atomic mass unit [u] or Dalton [Da] is used for calculation. The atomic mass is the 12th part of the mass of a carbon isotope ^{12}C, which corresponds in close approximation to the mass of one hydrogen atom, so one can say the weight of a hydrogen atom is 1 u.

$$1\,\text{u} = 1\,\text{Da} = 1.660\,538\,921(73) \times 10^{-27}\,\text{kg},$$
$$1\,\text{u} = 1\,\text{Da} = 1.660\,538\,921(73) \times 10^{-24}\,\text{g}. \tag{1.3}$$

Equation (1.4) shows the Avogadro constant N_A, which is simply the number of particles in one mole.

$$N_A = 6.022\,141\,29(27) \times 10^{23}\,\text{mol}^{-1}. \tag{1.4}$$

Multiplying 1 u with the Avogadro constant delivers a very good approximation of $1\,\text{g}\,\text{mol}^{-1}$. Since the weight of one hydrogen atom can be assumed to be 1 u, Eq. (1.5) shows that the atomic weight of an element can be easily derived by taking the same absolute value.

$$1\,\text{u} \cdot N_A = 0.999\,999\,999\,806\,148 \approx 1\,\text{g}\,\text{mol}^{-1}. \tag{1.5}$$

For example, the lithium element, which has 6.94 u, and by approximation, also $6.94\,\text{g}\,\text{mol}^{-1}$.

Table 1.1 Specific capacity of selected elements and corresponding redox reaction [1].

Element	Redox Reaction			Specific Capacity [mAhg^{-1}]
Hydrogen	$2\,H^+$	$\rightleftarrows$	$H_2(g)$	26591.32
Beryllium	Be^{2+}	$\rightleftarrows$	$Be(s)$	5947.80
Carbon Dioxide	$CO_2(g) + 2\,H^+$	$\rightleftarrows$	$CO(g) + H_2O$	4462.81
Lithium	Li^+	$\rightleftarrows$	Li	3861.32
Silicon	$Si(s) + 4\,H^+$	$\rightleftarrows$	$SiH_4(g)$	3817.05
Oxygen	$O_2 + 4\,H^+$	$\rightleftarrows$	$2\,H_2O$	3350.38
Aluminium	Al^{3+}	$\rightleftarrows$	$Al(s)$	2979.92
Phosphorus	$P + 3\,H^+$	$\rightleftarrows$	$PH_3(g)$	2595.86
Magnesium	Mg^{2+}	$\rightleftarrows$	$Mg(s)$	2205.42
Sulfur	$SO_2(aq) + 4\,H^+$	$\rightleftarrows$	$S(s) + 2\,H_2O$	1671.69
Iron	Fe^{3+}	$\rightleftarrows$	Fe	1439.77
Fluorine	$F_2(g)$	$\rightleftarrows$	$2\,F^-$	1410.75
Calcium	Ca^{2+}	$\rightleftarrows$	$Ca(s)$	1337.46
Boron Acid	$H_3BO_3 + 3\,H^+$	$\rightleftarrows$	$B(s) + 3\,H_2O$	1300.81
Sodium	Na^+	$\rightleftarrows$	Na	1165.78
Zinc	Zn^{2+}	$\rightleftarrows$	$Zn(s)$	819.87
Chlorine	$Cl_2(g)$	$\rightleftarrows$	$2\,Cl^-$	755.97
Cadmium	Cd^{2+}	$\rightleftarrows$	$Cd(s)$	476.85
Lead	Pb^{2+}	$\rightleftarrows$	$Pb(s)$	258.70

Table 1.1 summarizes the specific capacity of common elements — some already in use in commercialized batteries, while others are discussed as potential promising candidates or as examples why their high capacities cannot be transferred to modern batteries. The full list of all these elements and their positions in the electrochemical series is collected in Appendix A.

Because most electrochemical redox reactions that can be efficiently employed in electrochemical cells to store energy need only the transfer of one or two electrons, high values of specific capacities can be consequently found in the upper left and right rows of the periodic table. Therefore, as a rule of thumb, the specific capacity depends mostly on the molar mass of an element. The lightest elements, with low molar masses are located in the first rows of the periodic table. For example, the electrochemical redox reaction of lithium, which generally transfers one electron only, results in a high specific capacity value due to its low molar mass. However, the specific capacity of silicon results in more or less the same value though its molar mass is four times higher than the molar mass of lithium because it uses four electrons in the redox reaction process. The same holds for aluminium with a three-electron transfer. In this regard aluminium and silicon are exceptions since no other elements can compensate their higher molar mass by electron transfer in such an efficient way. Lithium and silicon have very high specific capacities, so both could be, at this point, potential electrode materials for secondary battery systems.

Besides capacity, the voltage, as compared to hydrogen, strongly affects the energy density in the materials vying to be the best choice for cell components in a secondary battery system. The energy density represents how much energy can be stored in the cell and sums up the absolute amount of the voltages of the two single electrodes, as compared to a reference, usually hydrogen or in case of Li-based cells, Li. This means that the voltage and the specific electrode capacities have to be always considered.

But, is it also feasible that the gravimetric energy density E^m of an element or compound will be able to provide a deeper insight than its specific capacity? The gravimetric energy density of an element shows how much energy per kilogram or gram can be stored, while its volumetric energy density shows how much energy per litre can be stored. In this regard these are the most important values to differentiate potential candidates for high-energy cells. To calculate the gravimetric energy density E^m of a single element or compound, the specific capacity C^m is multiplied with the electrochemical series voltage U^0, as shown in Eq. (1.6).

$$E^m = C^m \cdot U^0 = \frac{z \cdot F}{M_{AM}} \cdot U^0. \tag{1.6}$$

In the following, the values for the electrochemical series voltages are based on the hydrogen level. This means that the electrochemical redox reaction of hydrogen has the voltage of $U^0 = 0\,\text{V}$. Thus, one has to be careful with this value. However, for the choice of battery materials it makes some sense since hydrogen represents the middle ($\sim 3\,\text{V}$) of the reduction direction, and oxidation can be achieved. This also means that one single electrochemical cell can never exceed 5.9 V, which is the combination voltage of lithium and fluorine!

The series voltages of many relevant elements are listed in Appendix B. Since there are positive and negative values for the series voltages, it formally results in positive and negative gravimetric energy densities per element using hydrogen as a reference.

The values of the gravimetric energy densities of each element can thus be put into a reliable order employing the electrochemical series, using hydrogen as a reference as shown in Fig. 1.1. Only a selection of the elements is mapped

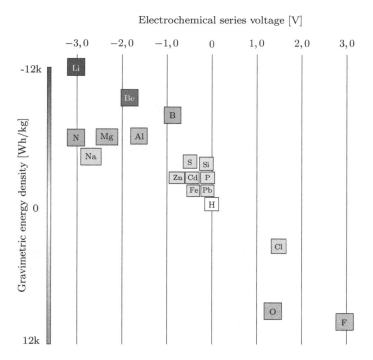

Fig. 1.1 Gravimetric energy density of selected elements per element, using hydrogen (H) as reference.

in this diagram; a more detailed table and calculated values are available in Appendix C.

The unit of the gravimetric energy density in Fig. 1.1 is Whkg^{-1}. The color intensity of the element symbols in Fig. 1.1 illustrates schematically higher values of the gravimetric energy density; the value is indicated on the ordinate.

Figure 1.1 indicates, as a feasible rule of thumb, that the elements with a high gravimetric energy density are those that have the highest and lowest electrochemical series voltage, using hydrogen as a reference. The hydrogen element symbol (H) is located in the middle of the diagram in Fig. 1.1 and is the reference point in this electrochemical series.

The material with the highest specific energy is the element lithium (Li) with 11 738 Whkg^{-1}. The element silicon (Si) which has more or less the same specific capacity as lithium, has only 534 Whkg^{-1}. This shows the big impact of the electrochemical series voltage to the gravimetric energy density value of a single element. Lithium has a high absolute value of electrochemical series voltage as compared to hydrogen, but silicon is close to the reference voltage of hydrogen. Consequently, this results in a huge difference between these two elements though they have almost the same specific capacity value. Silicon cannot even store half of the energy compared to lithium. In this regard, lithium is a very promising candidate for secondary battery systems.

However, this view is still simplified because in the end, two electrodes have to be combined, and though lead (Pb) may be a poor candidate following the above view, lead acid batteries are still widely commercialized, with Al-air primary batteries being widely researched because of its attractive specific capacity in the use of such cells.

Since the most popular field of application for secondary batteries is electromobility, the most important characteristic is volumetric energy density. This value shows how much energy can be stored in one liter of the element. Due to late progress in the development of Li-ion batteries, the volumetric energy density overrules the gravimetric one. For example, a vehicle not only needs to provide space for passengers and load, it must also look attractive.

To calculate the volumetric energy density for a single element, the molar mass needs to be divided by the density ρ of the element, as shown in Eq. (1.7).

$$E^V = \frac{z \cdot F \cdot U^0}{\frac{M_{AM}}{\rho}}. \tag{1.7}$$

The density of the elements can be easily drawn from the periodic table [2] and has the unit $g\,cm^{-3}$. From Eq. (1.7), the unit of the volumetric energy density is WhL^{-1}. Figure 1.2 shows selected elements ranked by their volumetric energy density. The full list of elements and their calculcated volumetric energy density is shown in Appendix D. Like Fig. 1.1, they are ordered according to their electrochemical series voltage, with hydrogen as a reference. In Fig. 1.2, there is no longer the mapping of positive and negative values. The volumetric energy density shows no trend to the series voltage, so it is more practical to work with absolute values. The color and the size of their symbols should indicate their volumetric energy density. The meaning of the colors is integrated in the ordinate of Fig. 1.2.

Here beryllium (Be) has the highest value of volumetric energy density. This means, at least theoretically, the most energy can be stored in one liter of this element, thus potentially leading to one of the smallest secondary batteries.

Lithium (Li), which was the ideal element as suggested earlier shows only an energy density of $6280\,WhL^{-1}$, and is located towards the middle

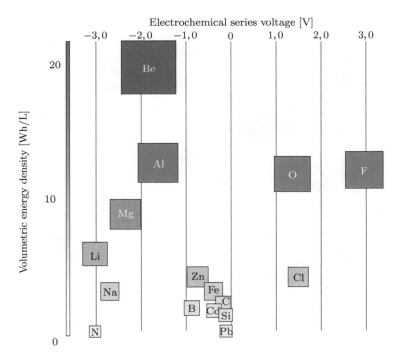

Fig. 1.2 Volumetric energy density of selected elements per element, using hydrogen as reference.

of the diagram in Fig. 1.2. It has even less volumetric energy density than aluminium (Al), magnesium (Mg) or oxygen (O), but still more than silicon (Si). The volumetric energy density of aluminium is almost twice compared to lithium, but its specific capacity, as shown in Table 1.1, is only about three-quarter of lithium and silicon.

Also, the volumetric energy density of aluminium is much higher than the value for silicon. As a result, lithium is not the only metal material to be considered for secondary battery systems.

To provide a suitable overview over potential elements of interest for secondary batteries, Fig. 1.3 compares the selected elements mentioned in this chapter and tries to add additional meaningful selection criteria.

According to Figs. 1.1 and 1.2, these elements were ordered by their electrochemical series voltage using hydrogen as reference. The ordinate in Fig. 1.3 is the specific capacity according to the values of Table 1.1. The color indicates the specific energy density as in Fig. 1.1 and the size of the element symbols indicates the value of the volumetric energy density derived from Fig. 1.2.

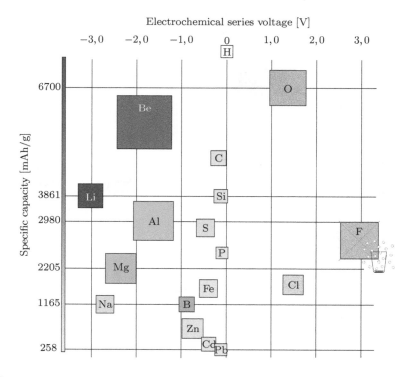

Fig. 1.3 Most suitable elements for high-energy density electrochemical cells.

After the main cell characteristics mentioned above were combined to visualize the best compromise for cell elements, lithium turned out to be one of the most suitable elements due to its lightweightedness (only hydrogen and helium exceed lithium in this discipline) and attractive voltage, as compared to hydrogen.

The overview of the most suitable materials for battery systems in Fig. 1.3 shows that though a single element can be an ideal candidate for high-energy density electrochemical cells, it may nevertheless not be recommended for a variety of reasons (e.g., beryllium is toxic while fluorine is gaseous) that renders them unsuitable for use in batteries.

After omitting these two elements, Fig. 1.3 summarizes the remaining high potential elements. Since there cannot be more than "one and only" elements for negative and positive electrodes, further shortlisting is required. As explained, though lithium is one of the top materials, magnesium and aluminium are also good candidates since they compensate for their higher weight by the transfer of two and three electrons per atom, respectively. Lithium, on the other hand, transfers only one electron per atom. Though

Fundamental Aspects of Achievable Energy Densities in Electrochemical Cells 9

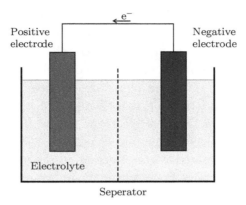

Fig. 1.4 Schematic set-up of an electrochemical cell.

silicon transfers four electrons per atom, it does not easily satisfy the requirements for a top single cell material, since voltage, if hydrogen is taken as a reference, is not very attractive any more. Thus as shown in Fig. 1.3, it may be on the same specific capacity value as lithium, but its color and symbol in the diagram are much lighter and smaller, respectively, as compared to lithium.

After having identified appropriate materials for electrochemical cells it is appropriate to deal with the basic structure of electrochemical cells. Every electrochemical cell is based on the structure shown in Fig. 1.4.

To build an efficient battery system it is necessary to identify two suitable electrodes. The positive and negative electrodes are built up by using one (or several) of the elements mentioned earlier on in this chapter. An electrochemical cell is defined by two half-cells, an optional separator that separates these two half-cells by a very short distance, and an electrolyte that provides efficient ion transfer between the half-cells.

To close the electric circuit, these two electrodes need an external connection for the electrons that electrical work can be supplied by the cell. Thus, an electrode needs to be a mixed conductor for ions and electrons. The positive electrode is the half-cell usually with a positive potential towards the hydrogen normal potential of the electrochemical series voltage. The negative electrode is commonly defined with a negative potential towards the hydrogen normal potential. However, it is the absolute voltage difference that counts for in the end and if the specific electrode capacity is very attractive, or for some other reason comes to the foreground (e.g., costs), two materials with negative or positive potential, using hydrogen as a reference, may be combined. An example for the latter is the lead-acid battery. Each electrode is one half-cell contributing to a whole electrochemical redox reaction.

There are two kinds of electrodes, distinctly differing in their way of behaving. First, the classic metal and metal compound (e.g., oxide) electrode where the active material of the electrode is chemically transferred into another compound, which usually causes ultimate mechanical stress upon deep cycling. The second kind of electrode is the intercalation electrode, where the active material shows how stoichiometric widths within atoms that can be inserted and extracted without changing the structural sublattice of the electrode, which is a huge benefit over the above-mentioned conventional electrodes.

Most of the electrodes in Li-ion cells are intercalation electrodes, for example, the lithium graphite LiC_6 electrode. A schematic view of a lithium-ion cell with two intercalation electrodes is shown in Fig. 1.5. The lithium-ions intercalate and deintercalate into and out of the structure of the host materials. This is like a sponge taking up and releasing water several thousand times! The benefit of such a process becomes obvious.

However, this principle is paid by incorporating additional elements that increase weight and diminish the specific electrode capacities.

Calculating the theoretical value of the specific capacity of a typical positive electrode employed in Li-ion cells, for example, lithium-nickel-cobalt-manganese oxide, $LiNi_{1/3}Co_{1/3}Mn_{1/3}O_2$, shows that the high specific capacity of lithium cannot be equally preserved on the positive side. Equation (1.8)

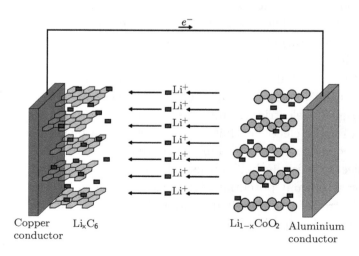

Fig. 1.5 Schematic structure of intercalation electrodes, for example, the $LiC_6/LiCoO_2$ cell.

shows how to calculate the value of $C^m{}_{\text{pos\_electrode}}$; thereby the lithium is entirely intercalated in the positive electrode. This equation premised on Eq. (1.2), which calculates the specific capacity of an element. For the electrode compound the molar mass M of the components is aggregated.

$$C^m{}_{\text{pos\_electrode}} = \frac{z \cdot F}{M_{\text{Li}} + \frac{1}{3} \cdot M_{\text{Ni}} + \frac{1}{3} \cdot M_{\text{Co}} + \frac{1}{3} \cdot M_{\text{Mn}} + 2 \cdot M_{\text{O}}},$$
$$= 1000.28 \, \text{Asg}^{-1} = 277.86 \, \text{mAhg}^{-1}. \quad (1.8)$$

The positive electrode $\text{LiNi}_{1/3}\text{Co}_{1/3}\text{Mn}_{1/3}\text{O}_2$ is one of the most commonly used for common lithium-ion cells. For negative electrodes, the specific capacity $C^m{}_{\text{neg\_electrode}}$ of the lithium graphite intercalation electrode, LiC_6, is calculated by Eq. (1.9), which also based on Eq. (1.2).

$$C^m{}_{\text{neg\_electrode}} = \frac{z \cdot F}{M_{\text{Li}} + 6 \cdot M_{\text{C}}},$$
$$= 1221.2 \, \text{Asg}^{-1} = 339.24 \, \text{mAhg}^{-1}. \quad (1.9)$$

This calculation shows that the intercalation of lithium into a structure like graphite results in a lower specific capacity of the electrode. A metal electrode of lithium, however, would provide the full specific capacity and specific energy of the lithium element, assuming that the full active material can be used for charging/discharging, which is still not possible in practical applications. A selection of lithium electrodes and intercalation and conversion electrodes is shown in Fig. 1.6. The theoretical value of the specific capacity is located in the middle of the diagram and the practical value, which can be achieved in a usable cell (within the regime where the sublattice is stable, for example, in case of intercalation electrodes), is shown at the edges.

The most common negative lithium-ion electrode is lithium graphite, LiC_6, which is located at the bottom quandrant of the left diagram in Fig. 1.6. It nearly reaches the theoretical specific capacity of practical use and has a long cycle life. The negative electrode in the right quandrant of the left square is the intercalation compound, $\text{Li}_4\text{Ti}_5\text{O}_{12}$, which is excellent at cycling but has a low capacity. The lithium metal electrode is the negative electrode with the highest specific capacity but it is still at a research phase, with the hope of producing an electrode with at least good cycling performance and without dendrite growth. The fourth negative electrode in Fig. 1.6 is the lithium silicon electrode, which has a good specific capacity with regard to

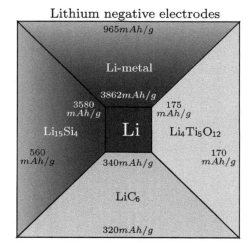 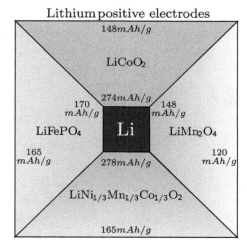

Fig. 1.6 Selected lithium electrodes to compare theoretical and practical specific capacity values in half-cells.

the commonly used electrodes but has a high volumetric change during the lithium intercalation [7].

The most common positive electrodes of lithium-ion cells are $LiNi_{1/3}Co_{1/3}Mn_{1/3}O_2$ and $LiCoO_2$, but they are both expensive due to the presence of cobalt. A low-cost positive electrode alternative is $LiMn_2O_4$; however this material shows a low specific capacity and limited cycle life, especially at elevated temperatures, which is an effect of the Manganese dissolution in the liquid electrolyte. Another electrode, $LiFePO_4$, has excellent safety performance but exhibits very low specific capacity [8], and most importantly, due to its position in the electrochemical series of elements it has a lower voltage in a Li-ion cell, which reduces its energy density.

To summarize Fig. 1.6, most of these theoretical specific capacities mentioned above cannot yet be implemented to a real cell, especially since the positive intercalation electrodes for Li-ion cells are all still at very low values. However, omitting the intercalation principle would result in cells with unacceptable lifetime expectations.

The electrochemical series voltages U^0 of the cell compounds are of course different to that of the lithium metal electrode. To obtain a high-energy density it is recommended that the series voltages of the electrodes be spread. Figure 1.7 shows the mentioned lithium electrodes with the related series voltage referenced to the Li/Li^+ redox reaction and the spreading potential of different compounds. However, this must be matched with the specific capacities to obtain a maximum achievable energy density.

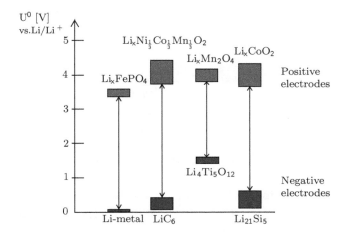

Fig. 1.7 Selected lithium electrodes related to the electrochemical series voltage, in reference to Li/Li$^+$.

On the vertical axis (reference voltage) where the lithium metal electrode is depicted, U^0 of Li/Li$^+$ is 0 V. As the intercalation compound Li$_4$Ti$_5$O$_{12}$ electrode has a higher voltage, as compared to the Li value, comparatively to the other negative electrodes, so battery systems with this negative electrode tend to have lower specific energy values. The LiNi$_{1/3}$Co$_{1/3}$Mn$_{1/3}$O$_2$ compound has a higher voltage, as opposed to Li compared to LiFePO$_4$, but accumulates worse cycling, aging and safety behavior.

This practical set-up of such electrodes can never reach the theoretical value of their specific capacity, because the lithium cannot be completely squeezed out of the electrode structure without making the structure heavily unstable. But if a dendrite build-up stopper during the cycles of these lithium-ion systems is somehow implemented (since lithium can never be completely separated from the structure), they will never appear metallic if these electrodes are properly operated.

Thus, in order to calculate the characteristic energy density values of a Li-ion cell, it is necessary to know how much of the lithium must stay intercalated in an electrode to guarantee a stable structure. Using the positive electrode lithium-nickel-cobalt-manganese-oxide as an example, 0.4 of the lithium needs to stay intercalated.

This results in two different states of the cell — the discharged and the charged one. In the charged cell there are 0.4 parts of lithium still intercalated in the nickel-cobalt-manganese-oxide structure, so that the specific capacity values of the electrodes can be calculated for the charged and discharged states of a cell. Equation (1.10) shows the calculation of the specific capacity of the positive electrode in the charged state.

$$C^m_{\text{pos\_electrode\_charged}} = \frac{z \cdot F}{0.4 \cdot M_{\text{Li}} + \frac{1}{3} \cdot M_{\text{Ni}} + \frac{1}{3} \cdot M_{\text{Co}} + \frac{1}{3} \cdot M_{\text{Mn}} + 2 \cdot M_{\text{O}}}$$
$$= 631.17 \, \text{Asg}^{-1} = 175.32 \, \text{mAhg}^{-1}. \qquad (1.10)$$

The negative electrode can uptake one Li and becomes LiC_6, which results in a specific capacity of $339.24 \, \text{mAhg}^{-1}$. The positive and negative electrodes together form an electrochemical cell. The specific capacity of the charged full-cell is calculated on the basis of the specific capacities of each electrode, as shown in Eq. (1.11).

$$\frac{1}{C^m_{\text{cell\_charged}}} = \frac{1}{C^m_{\text{pos\_electrode\_charged}}} + \frac{1}{C^m_{\text{neg\_electrode\_charged}}}. \qquad (1.11)$$

So, the example $0.6\text{Li}_6/\text{Li}_{0.4}\text{Ni}_{1/3}\text{Co}_{1/3}\text{Mn}_{1/3}\text{O}_2$ results in the theoretical cell specific capacity to $115.59 \, \text{mAhg}^{-1}$. Taking into consideration the nominal voltage U_N of a fully charged lithium-ion cell, it is also possible to calculate its specific energies (gravimetric/volumetric). Additionally, it is necessary to know the part of lithium which can be transferred in the redox reaction without compromising the structural integrity of the electrodes and the active mass of both electrodes. The calculation of the specific energy based on Eq. (1.6), with nominal cell voltage as $3.8 \, \text{V}$, the gravimetric energy density of a charged cell using the previous example results in $439.23 \, \text{Whg}^{-1}$.

Using the same positive electrode, but employing metallic lithium instead of LiC_6 as a negative electrode, the gravimetric energy density rises to $635.66 \, \text{Whkg}^{-1}$. This shows that the positive intercalation-based electrode is the bottleneck. An overview of the theoretical gravimetric energy density E^m of several electrochemical systems is shown in Fig. 1.8.

Figure 1.8 also shows the commonly used and proposed secondary battery systems — the not Li-based ones in yellow and the potential post lithium systems in light violet. The previously discussed example of a lithium-ion system is marked red and can easily be compared to the already mentioned lithium system in Table 1.2 with a commonly used positive electrode and lithium metal as a negative electrode. The best benefit would be a positive

Fundamental Aspects of Achievable Energy Densities in Electrochemical Cells 15

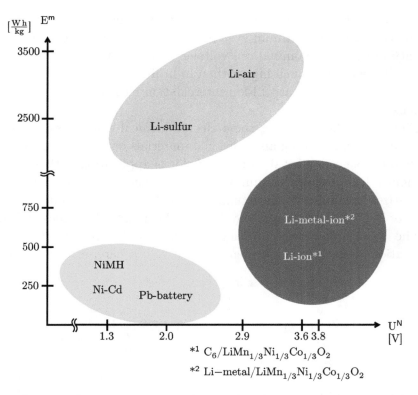

Fig. 1.8 Theoretical gravimetric energy densities based on pure active material electrode calculation (no other components such as separator, current collectors, conductive agents, etc., are included).

Table 1.2 Values for Fig. 1.8.

Battery system	Series voltage [V]	Theoretical specific energy $[\frac{Wh}{kg}]$
Pb-battery	2.0	161
NiCd	1.3	210
NiMH	1.3	359
Li-ion*1	3.6	439
Li-metal-ion*2	3.8	635
Li-sulfur	2.1	2447
Li-air	2.9	3378
Li-flourine	6.0	4873

electrode material, which includes an increase in nominal cell voltage and specific capacity. Post lithium systems like lithium air increase the specific capacity but not the nominal cell voltage. However, the theoretical value of the specific energy is much higher than in lithium-ion systems. This demonstrates again how standard Li-ion intercalation electrodes can lead to huge bottlenecks.

The last challenge facing these electrochemical cells is comparing their practical values, including all additional materials on cell and battery levels, as this value is needed for all potential fields of application of these batteries.

Figure 1.9 provides a summary of this effort on the cell level and summarizes how full-cells are actually assessed. Table 1.3 includes the specific values of Fig. 1.9. In the same way, Table 1.4 contains the values of Fig. 1.10.

The huge differences in Figs. 1.9 and 1.10 need further detailed explanation and their consequences have to be carefully noted: For the

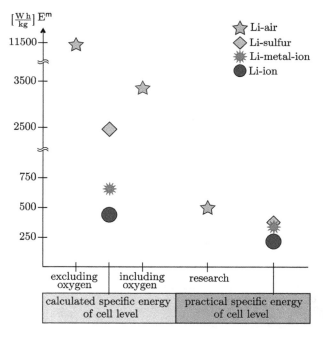

Fig. 1.9 Comparison of achievable gravimetric energy densities in full-cells (projected/practical) versus electrode calculation. Solid Li-ion means Li-metal-ion. Battery means assembled cells.

Table 1.3 Values for Fig. 1.9 [6], [9], [10], [11].

Battery system	Cell level Calculated gravimetric energy density $\frac{Wh}{kg}$	Cell level Practical gravimetric energy density $\frac{Wh}{kg}$
Li-air: $(2\,Li + O_2 \longrightarrow Li_2O_2)$ excluding oxygen	11680	–
Li-air: $(2\,Li + O_2 \longrightarrow Li_2O_2)$ including oxygen	3460	362 (state of research)
8Li2S)	2500	350
Li-metal-ion: $(Li-metal/LiMn_{1/3}Ni_{1/3}Co_{1/3}O_2)$	635	≈ 318
Li-ion: $(LiC_6/LiMn_{1/3}Ni_{1/3}Co_{1/3}O_2)$	439	≈ 220

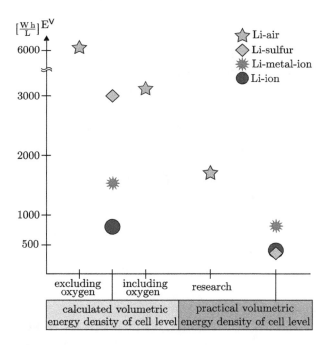

Fig. 1.10 Comparison of achievable volumetric energy densities in full-cells (projected/practical) versus electrode calculation. Solid Li-ion means Li-metal-ion. Battery means assembled cells.

Table 1.4 Values for Fig. 1.10 [9] [11].

Battery system	Cell level Calculated volumetric energy density $\left[\frac{Wh}{L}\right]$	Cell level Practical volumetric energy density $\left[\frac{Wh}{L}\right]$
Li-air: $(2\,Li + O_2 \longrightarrow Li_2O_2)$ excluding oxygen	6197	–
Li-air: $(2\,Li + O_2 \longrightarrow Li_2O_2)$ including oxygen	3323	1000 (state of research)
Li-sulfur: $(S_8 + 16\,Li \longleftrightarrow 8\,Li_2S)$	2967	300
Li-metal-Ion: $(Li\text{-metal}/Mn_{1/3}Ni_{1/3}Co_{1/3}O_2)$	1485	≈ 742
Li-Ion: $(LiC_6/Mn_{1/3}Ni_{1/3}Co_{1/3}O_2)$	798	≈ 399

calculated theoretic value of the specific energy density, it is necessary to differentiate the lithium air system in states that either include or exclude oxygen. For an open system, because oxygen for the reaction comes from the ambient air, it is not a full part of the battery system! This results in very high specific energy for lithium air without oxygen (fully charged), as shown in Fig. 1.9. Nonetheless, the weight of oxygen must somehow be included. In this regard there is a second calculated specific energy value for lithium air systems in the fully discharged state. This is the consequence of open systems. For example, the lithium sulphur and lithium-ion systems are, however, always closed, so there is only one calculated value of specific energy density. But, the decrease from calculated value to projected value is much bigger for the lithium sulphur system than it is for the lithium-ion system. This is strongly dependent on how much aid is needed to access an element or compound, electrochemically and reversibility. For example, sulphur needs a large amount of conducting agent, while Li-metal cells need additional reserve lithium to stay rechargeable.

To summarize, though some systems offer fantastic electrode calculation (only active material) values on paper, there seems to be a cumulative effect on the cell level where all systems come quite close to advanced Li-ion. Because of this, future improvements to actual values will be about a factor of 1.5 to 2 for energy densities. This makes modern battery engineering inevitable.

Annex

A. Specific capacity of each element

Table A.1 The specific capacity of single elements related to the redox reaction [1].

Element	Redox reaction			Specific capacity [mAhg^{-1}]
Hydrogen	$2\,H^+$	$\rightleftarrows$	$H_2(g)$	26591.32
Beryllium	Be^{2+}	$\rightleftarrows$	$Be(s)$	5947.80
Carbon Dioxide	$CO_2(g) + 2\,H^+$	$\rightleftarrows$	$CO(g) + H_2O$	4462.81
Lithium	Li^+	$\rightleftarrows$	Li	3861.32
Silicon	$Si(s) + 4\,H^+$	$\rightleftarrows$	$SiH_4(g)$	3817.05
Oxygen	$O_2 + 4\,H^+$	$\rightleftarrows$	$2\,H_2O$	3350.38
Aluminium	Al^{3+}	$\rightleftarrows$	$Al(s)$	2979.92
Phosphorus	$P + 3\,H^+$	$\rightleftarrows$	$PH_3(g)$	2595.86
Magnesium	Mg^{2+}	$\rightleftarrows$	$Mg(s)$	2205.42
Titanium	Ti^{3+}	$\rightleftarrows$	$Ti(s)$	1679.74
Sulfur	$SO_2(aq) + 4\,H^+$	$\rightleftarrows$	$S(s) + 2\,H_2O$	1671.69
Chromium	Cr^{3+}	$\rightleftarrows$	Cr	1546.35
Germanium	$GeO(s) + 2\,H^+$	$\rightleftarrows$	$Ge(s) + H_2O$	1475.85
Iron	Fe^{3+}	$\rightleftarrows$	Fe	1439.77
Fluorine	$F_2(g)$	$\rightleftarrows$	$2\,F^-$	1410.75
Calcium	Ca^{2+}	$\rightleftarrows$	$Ca(s)$	1337.46
Boron Acid	$H_3BO_3 + 3\,H^+$	$\rightleftarrows$	$B(s) + 3\,H_2O$	1300.81
Zirconium	$ZrO(OH)_2 + H_2O$	$\rightleftarrows$	$Zr + 4\,OH^-$	1175.19
Sodium	Na^+	$\rightleftarrows$	Na	1165.78
Gallium	Ga^{3+}	$\rightleftarrows$	Ga	1153.19
Arsenic	$H_3AsO_3 + 3\,H^+$	$\rightleftarrows$	$As(s) + 3\,H_2O$	1073.17
Vanadium	V^{2+}	$\rightleftarrows$	$V(s)$	1052.23
Manganese	Mn^{2+}	$\rightleftarrows$	$Mn(s)$	975.70
Nickel	Ni^{2+}	$\rightleftarrows$	$Ni(s)$	913.27
Cobalt	Co^{2+}	$\rightleftarrows$	$Co(s)$	909.55
Yttrium	Y^{3+}	$\rightleftarrows$	$Y(s)$	904.37
Niobium	Nb^{3+}	$\rightleftarrows$	$Nb(s)$	865.44
Molybdenum	$MoO_2(s) + 4\,H^+$	$\rightleftarrows$	$Mo(s) + 2\,H_2O$	837.89
Zinc	Zn^{2+}	$\rightleftarrows$	$Zn(s)$	819.87
Chlorine	$Cl_2(g)$	$\rightleftarrows$	$2\,Cl^-$	755.97

(*Continued*)

Table A.1 (*Continued*)

Element	Redox Reaction			Specific Capacity [mAhg^{-1}]
Indium	In^{3+}	$\rightleftarrows$	$In(s)$	700.26
Potassium	K^+	$\rightleftarrows$	$K(s)$	685.49
Selenium	$Se(s) + 2H^+$	$\rightleftarrows$	$H_2Se(g)$	678.86
Antimony	$SbO^+ + 2H^+$	$\rightleftarrows$	$Sb(s) + H_2O$	660.35
Strontium	Sr^{2+}	$\rightleftarrows$	$Sr(s)$	611.76
Tungsten	$WO_2(s) + 4H^+$	$\rightleftarrows$	$W(s) + 2H_2O$	583.15
Lanthanum	La^{3+}	$\rightleftarrows$	$La(s)$	578.82
Technetium	$TcO_2 + 4H^+$	$\rightleftarrows$	$Tc + 4H_2O$	547.49
Terbium	Tb^{3+}	$\rightleftarrows$	Tb	505.91
Palladium	Pd^{2+}	$\rightleftarrows$	$Pd(s)$	503.69
Iron	Fe^{3+}	$\rightleftarrows$	Fe^{2+}	479.92
Cadmium	Cd^{2+}	$\rightleftarrows$	$Cd(s)$	476.85
Tin	Sn^{2+}	$\rightleftarrows$	$Sn(s)$	451.54
Tantalum	Ta^{3+}	$\rightleftarrows$	$Ta(s)$	444.34
Rhenium	Re^{3+}	$\rightleftarrows$	$Re(s)$	431.79
Copper	Cu^+	$\rightleftarrows$	$Cu(s)$	421.76
Tellurium	$Te(s)$	$\rightleftarrows$	Te^{2-}	420.08
Iridium	$Ir^{3+} + 2H_2O$	$\rightleftarrows$	Ir	418.29
Gold	Au^{3+}	$\rightleftarrows$	$Au(s)$	408.21
Barium	Ba^{2+}	$\rightleftarrows$	$Ba(s)$	390.32
Bismuth	Bi^{3+}	$\rightleftarrows$	$Bi(s)$	354.16
Actinium	Ac^{3+}	$\rightleftarrows$	$Ac(s)$	354.16
Europium	Eu^{2+}	$\rightleftarrows$	$Eu(s)$	352.74
Uranium	U^{3+}	$\rightleftarrows$	$U(s)$	337.79
Bromine	Br_2	$\rightleftarrows$	$2Br^-$	335.42
Rubidium	Rb^+	$\rightleftarrows$	$Rb(s)$	313.58
Platinum	Pt^{2+}	$\rightleftarrows$	$Pt(s)$	274.77
Mercury	Hg^{2+}	$\rightleftarrows$	Hg	267.23
Lead	Pb^{2+}	$\rightleftarrows$	$Pb(s)$	258.70
Silver	Ag^+	$\rightleftarrows$	$Ag(s)$	248.46
Radium	Ra^{2+}	$\rightleftarrows$	$Ra(s)$	237.15
Iodine	$I_2(s)$	$\rightleftarrows$	$2I^-$	211.20
Caesium	Cs^+	$\rightleftarrows$	Cs	201.65
Cerium	Ce^{4+}	$\rightleftarrows$	Ce^{3+}	191.27
Thallium	Tl^+	$\rightleftarrows$	$Tl(s)$	131.14

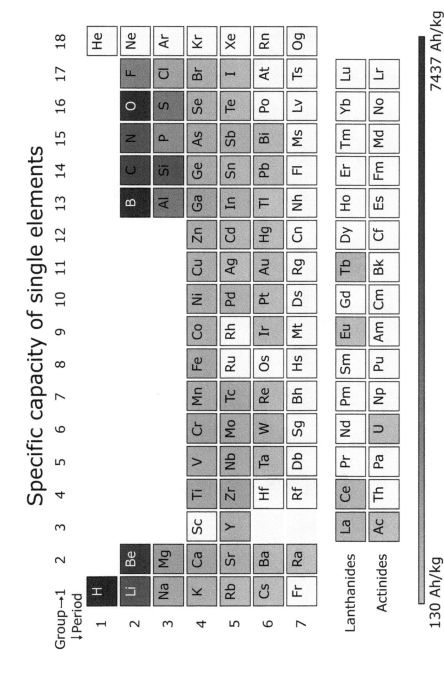

Fig. A.1 Specific capacity of each element of the periodic table.

B. Series voltage of each element

Table B.1 The eletrochemical series voltage related to the redox reaction [1].

Element	Redox reaction			Series voltage [V]
Lithium	Li^+	$\rightleftarrows$	Li	−3.04
Caesium	Cs^+	$\rightleftarrows$	Cs	−3.03
Rubidium	Rb^+	$\rightleftarrows$	Rb(s)	−2.98
Potassium	K^+	$\rightleftarrows$	K	−2.93
Barium	Ba^{2+}	$\rightleftarrows$	Ba(s)	−2.92
Strontium	Sr^{2+}	$\rightleftarrows$	Sr	−2.89
Calcium	Ca^{2+}	$\rightleftarrows$	Ca	−2.87
Europium	Eu^{2+}	$\rightleftarrows$	Eu(s)	−2.81
Radium	Ra^{2+}	$\rightleftarrows$	Ra(s)	−2.8
Sodium	Na^+	$\rightleftarrows$	Na	−2.71
Lanthanum	La^{3+}	$\rightleftarrows$	La(s)	−2.38
Yttrium	Y^{3+}	$\rightleftarrows$	Y	−2.37
Magnesium	Mg^{2+}	$\rightleftarrows$	Mg(s)	−2.36
Zirconium	$ZrO(OH)_2 + H_2O$	$\rightleftarrows$	$Zr + 4\,OH^-$	−2.36
Terbium	Tb^{3+}	$\rightleftarrows$	Tb	−2.31
Actinium	Ac^{3+}	$\rightleftarrows$	Ac(s)	−2.2
Beryllium	Be_2^+	$\rightleftarrows$	Be(s)	−1.85
Aluminium	Al_3^+	$\rightleftarrows$	Al(s)	−1.66
Uranium	U^{3+}	$\rightleftarrows$	U(s)	−1.66
Titanium	Ti^{3+}	$\rightleftarrows$	Ti(s)	−1.37
Manganese	Mn^{2+}	$\rightleftarrows$	Mn	−1.18
Tellurium	$Te(s)$	$\rightleftarrows$	Te^{2-}	−1.14
Vanadium	V^{2+}	$\rightleftarrows$	V	−1.13
Niobium	Nb^{3+}	$\rightleftarrows$	Nb	−1.1
Boron Acid	$H_3BO_3 + 3\,H^+$	$\rightleftarrows$	$B(s) + 3\,H_2O$	−0.89
Zinc	Zn^{2+}	$\rightleftarrows$	Zn	−0.76
Chromium	Cr^{3+}	$\rightleftarrows$	Cr	−0.74
Tantalum	Ta^{3+}	$\rightleftarrows$	Ta(s)	−0.6
Gallium	Ga^{3+}	$\rightleftarrows$	Ga	−0.53
Cadmium	Cd^{2+}	$\rightleftarrows$	Cd(s)	−0.4
Indium	In^{3+}	$\rightleftarrows$	In	−0.34
Thallium	Tl^+	$\rightleftarrows$	Tl(s)	−0.34

(*Continued*)

Table B.1 (*Continued*)

Element	Redox reaction			Series voltage [V]
Cobalt	Co^{2+}	$\rightleftarrows$	Co	−0.28
Nickel	Ni^{2+}	$\rightleftarrows$	Ni	−0.25
Molybdenum	$MoO_2(s) + 4\,H^+$	$\rightleftarrows$	$Mo(s) + 2\,H_2O$	−0.15
Silicon	$Si(s) + 4\,H^+$	$\rightleftarrows$	$SiH_4(g)$	−0.14
Tin	Sn^{2+}	$\rightleftarrows$	$Sn(s)$	−0.13
Lead	Pb^{2+}	$\rightleftarrows$	$Pb(s)$	−0.13
Tungsten	$WO_2(s) + 4\,H^+$	$\rightleftarrows$	$W(s) + 2\,H_2O$	−0.12
Phosphorus	$P + 3\,H^+$	$\rightleftarrows$	$PH_3(g)$	−0.11
Carbon Dioxide	$CO_2(g) + 2\,H^+$	$\rightleftarrows$	$CO(g) + H_2O$	−0.11
Selenium	$Se(s) + 2\,H^+$	$\rightleftarrows$	$H_2Se(g)$	−0.11
Iron	Fe^{3+}	$\rightleftarrows$	Fe	−0.04
Hydrogen	$2\,H^+$	$\rightleftarrows$	$H_2(g)$	0
Antimony	$SbO^+ + 2\,H^+$	$\rightleftarrows$	$Sb(s) + H_2O$	0.2
Arsenic	$H_3AsO_3 + 3\,H^+$	$\rightleftarrows$	$As(s) + 3\,H_2O$	0.24
Germanium	$GeO(s) + 2\,H^+$	$\rightleftarrows$	$Ge(s) + H_2O$	0.26
Technetium	$TcO_2 + 4\,H^+$	$\rightleftarrows$	$Tc + 4\,H_2O$	0.27
Rhenium	Re^{3+}	$\rightleftarrows$	$Re(s)$	0.3
Bismuth	Bi^{3+}	$\rightleftarrows$	$Bi(s)$	0.31
Sulfur	$SO_2(aq) + 4\,H^+$	$\rightleftarrows$	$S(s) + 2\,H_2O$	0.5
Copper	Cu^+	$\rightleftarrows$	$Cu(s)$	0.52
Iodine	$I_2(s)$	$\rightleftarrows$	$2\,I^-$	0.54
Iron	Fe^{3+}	$\rightleftarrows$	Fe^{2+}	0.77
Silver	Ag^+	$\rightleftarrows$	$Ag(s)$	0.8
Mercury	Hg^{2+}	$\rightleftarrows$	Hg	0.85
Palladium	Pd^{2+}	$\rightleftarrows$	$Pd(s)$	0.92
Bromine	Br_2	$\rightleftarrows$	$2\,Br^-$	1.07
Iridium	$Ir^{3+} + 2\,H_2O$	$\rightleftarrows$	Ir	1.16
Platinum	Pt^{2+}	$\rightleftarrows$	$Pt(s)$	1.19
Oxygen	$O_2 + 4\,H^+$	$\rightleftarrows$	$2\,H_2O$	1.23
Chlorine	$Cl_2(g)$	$\rightleftarrows$	$2\,Cl^-$	1.36
Cerium	Ce^{4+}	$\rightleftarrows$	Ce^{3+}	1.44
Gold	Au^{3+}	$\rightleftarrows$	$Au(s)$	1.52
Fluorine	$F_2(g)$	$\rightleftarrows$	$2\,F^-$	2.87

C. Specific energy of each element

Table C.1 The gravimetric energy density of single elements related to the redox reaction [1].

Element	Redox reaction			Gravimetric energy density [Whg^{-1}]
Lithium	Li$^+$	$\rightleftarrows$	Li	−11738.40
Beryllium	Be$_2{}^+$	$\rightleftarrows$	Be(s)	−11003.43
Magnesium	Mg^{2+}	$\rightleftarrows$	Mg(s)	−5209.21
Aluminium	Al$_3{}^+$	$\rightleftarrows$	Al(s)	−4946.66
Calcium	Ca^{2+}	$\rightleftarrows$	Ca	−3838.51
Sodium	Na$^+$	$\rightleftarrows$	Na	−3159.28
Zirconium	ZrO(OH)$_2$ + H$_2$O	$\rightleftarrows$	Zr + 4 OH$^-$	−2773.45
Titanium	Ti^{3+}	$\rightleftarrows$	Ti(s)	−2301.24
Yttrium	Y^{3+}	$\rightleftarrows$	Y	−2145.17
Potassium	K$^+$	$\rightleftarrows$	K	−2009.18
Strontium	Sr^{2+}	$\rightleftarrows$	Sr	−1767.39
Lanthanum	La^{3+}	$\rightleftarrows$	La(s)	−1377.02
Vanadium	V^{2+}	$\rightleftarrows$	V	−1189.02
Terbium	Tb^{3+}	$\rightleftarrows$	Tb	−1168.65
Boron Acid	H$_3$BO$_3$ + 3 H$^+$	$\rightleftarrows$	B(s) + 3 H$_2$O	−1157.72
Manganese	Mn^{2+}	$\rightleftarrows$	Mn	−1151.32
Chromium	Cr^{3+}	$\rightleftarrows$	Cr	−1144.30
Barium	Ba^{2+}	$\rightleftarrows$	Ba(s)	−1139.74
Europium	Eu^{2+}	$\rightleftarrows$	Eu(s)	−991.91
Niobium	Nb^{3+}	$\rightleftarrows$	Nb	−951.11
Rubidium	Rb$^+$	$\rightleftarrows$	Rb(s)	−934.48
Actinium	Ac^{3+}	$\rightleftarrows$	Ac(s)	−779.15
Radium	Ra^{2+}	$\rightleftarrows$	Ra(s)	−664.03
Zinc	Zn^{2+}	$\rightleftarrows$	Zn	−623.10
Gallium	Ga^{3+}	$\rightleftarrows$	Ga	−611.19
Caesium	Cs$^+$	$\rightleftarrows$	Cs	−610.19
Uranium	U^{3+}	$\rightleftarrows$	U(s)	−560.73
Silicon	Si(s) + 4 H$^+$	$\rightleftarrows$	SiH$_4$(g)	−534.39
Carbon Dioxide	CO$_2$(g) + 2 H$^+$	$\rightleftarrows$	CO(g) + H$_2$O	−490.91
Tellurium	Te(s)	$\rightleftarrows$	Te^{2-}	−480.16
Phosphorus	P + 3 H$^+$	$\rightleftarrows$	PH$_3$(g)	−288.14

(*Continued*)

Table C.1 (*Continued*)

Element	Redox reaction			Gravimetric energy density [Whg^{-1}]
Tantalum	Ta^{3+}	$\rightleftarrows$	$Ta(s)$	−266.61
Cobalt	Co^{2+}	$\rightleftarrows$	Co	−254.68
Indium	In^{3+}	$\rightleftarrows$	In	−238.09
Nickel	Ni^{2+}	$\rightleftarrows$	Ni	−228.32
Cadmium	Cd^{2+}	$\rightleftarrows$	$Cd(s)$	−190.74
Molybdenum	$MoO_2(s) + 4H^+$	$\rightleftarrows$	$Mo(s) + 2H_2O$	−125.68
Selenium	$Se(s) + 2H^+$	$\rightleftarrows$	$H_2Se(g)$	−74.67
Tungsten	$WO_2(s) + 4H^+$	$\rightleftarrows$	$W(s) + 2H_2O$	−69.98
Tin	Sn^{2+}	$\rightleftarrows$	$Sn(s)$	−58.70
Iron	Fe^{3+}	$\rightleftarrows$	Fe	−57.59
Thallium	Tl^+	$\rightleftarrows$	$Tl(s)$	−44.59
Lead	Pb^{2+}	$\rightleftarrows$	$Pb(s)$	−33.63
Hydrogen	$2H^+$	$\rightleftarrows$	$H_2(g)$	0
Iodine	$I_2(s)$	$\rightleftarrows$	$2I^-$	113.10
Bismuth	Bi^{3+}	$\rightleftarrows$	$Bi(s)$	118.50
Rhenium	Re^{3+}	$\rightleftarrows$	$Re(s)$	129.54
Antimony	$SbO^+ + 2H^+$	$\rightleftarrows$	$Sb(s) + H_2O$	132.07
Technetium	$TcO_2 + 4H^+$	$\rightleftarrows$	$Tc + 4H_2O$	148.92
Silver	Ag^+	$\rightleftarrows$	$Ag(s)$	198.77
Copper	Cu^+	$\rightleftarrows$	$Cu(s)$	219.32
Mercury	Hg^{2+}	$\rightleftarrows$	Hg	227.14
Arsenic	$H_3AsO_3 + 3H^+$	$\rightleftarrows$	$As(s) + 3H_2O$	257.56
Cerium	Ce^{4+}	$\rightleftarrows$	Ce^{3+}	275.43
Platinum	Pt^{2+}	$\rightleftarrows$	$Pt(s)$	326.43
Bromine	Br_2	$\rightleftarrows$	$2Br^-$	357.29
Iron	Fe^{3+}	$\rightleftarrows$	Fe^{2+}	369.54
Germanium	$GeO(s) + 2H^+$	$\rightleftarrows$	$Ge(s) + H_2O$	383.72
Palladium	Pd^{2+}	$\rightleftarrows$	$Pd(s)$	460.88
Iridium	$Ir^{3+} + 2H_2O$	$\rightleftarrows$	Ir	483.55
Gold	Au^{3+}	$\rightleftarrows$	$Au(s)$	620.47
Sulfur	$SO_2(aq) + 4H^+$	$\rightleftarrows$	$S(s) + 2H_2O$	835.85
Chlorine	$Cl_2(g)$	$\rightleftarrows$	$2Cl^-$	1028.12
Fluorine	$F_2(g)$	$\rightleftarrows$	$2F^-$	4048.85
Oxygen	$O_2 + 4H^+$	$\rightleftarrows$	$2H_2O$	4120.97

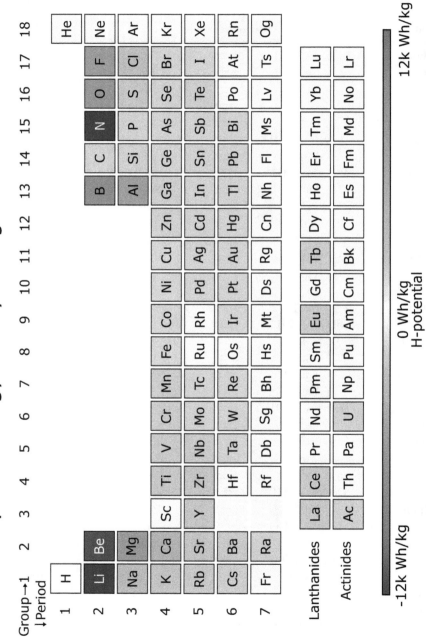

Fig. C.1 Specific energy density of each element of the periodic table.

D. Volumetric energy density of each element

Table D.1 The volumetric energy density of single elements related to the redox reaction [1].

Element	Redox reaction			Volumetric energy density [Wh/L]
Hydrogen	$2\,H^+$	$\rightleftarrows$	$H_2(g)$	0
Tin	Sn^{2+}	$\rightleftarrows$	$Sn(s)$	337.53
Selenium	$Se(s) + 2\,H^+$	$\rightleftarrows$	$H_2Se(g)$	359.11
Lead	Pb^{2+}	$\rightleftarrows$	$Pb(s)$	381.44
Iron	Fe^{3+}	$\rightleftarrows$	Fe	453.47
Phosphorus	$P + 3\,H^+$	$\rightleftarrows$	$PH_3(g)$	524.42
Thallium	Tl^+	$\rightleftarrows$	$Tl(s)$	531.91
Iodine	$I_2(s)$	$\rightleftarrows$	$2\,I^-$	557.57
Antimony	$SbO^+ + 2\,H^+$	$\rightleftarrows$	$Sb(s) + H_2O$	882.89
Bismuth	Bi^{3+}	$\rightleftarrows$	$Bi(s)$	1162.15
Caesium	Cs^+	$\rightleftarrows$	Cs	1177.68
Silicon	$Si(s) + 4\,H^+$	$\rightleftarrows$	$SiH_4(g)$	1244.91
Molybdenum	$MoO_2(s) + 4\,H^+$	$\rightleftarrows$	$Mo(s) + 2\,H_2O$	1292.03
Tungsten	$WO_2(s) + 4\,H^+$	$\rightleftarrows$	$W(s) + 2\,H_2O$	1350.57
Rubidium	Rb^+	$\rightleftarrows$	$Rb(s)$	1431.62
Bromine	Br_2	$\rightleftarrows$	$2\,Br^-$	1447.02
Arsenic	$H_3AsO_3 + 3\,H^+$	$\rightleftarrows$	$As(s) + 3\,H_2O$	1487.67
Cadmium	Cd^{2+}	$\rightleftarrows$	$Cd(s)$	1649.90
Technetium	$TcO_2 + 4\,H^+$	$\rightleftarrows$	$Tc + 4\,H_2O$	1712.53
Potassium	K^+	$\rightleftarrows$	K	1719.86
Sulfur	$SO_2(aq) + 4\,H^+$	$\rightleftarrows$	$S(s) + 2\,H_2O$	1721.84
Carbon Dioxide	$CO_2(g) + 2\,H^+$	$\rightleftarrows$	$CO(g) + H_2O$	1724.56
Boron Acid	$H_3BO_3 + 3\,H^+$	$\rightleftarrows$	$B(s) + 3\,H_2O$	1725.00
Indium	In^{3+}	$\rightleftarrows$	In	1740.43
Cerium	Ce^{4+}	$\rightleftarrows$	Ce^{3+}	1883.98
Copper	Cu^+	$\rightleftarrows$	$Cu(s)$	1956.31
Nickel	Ni^{2+}	$\rightleftarrows$	Ni	2033.86
Germanium	$GeO(s) + 2\,H^+$	$\rightleftarrows$	$Ge(s) + H_2O$	2042.54
Silver	Ag^+	$\rightleftarrows$	$Ag(s)$	2085.08
Chlorine	$Cl_2(g)$	$\rightleftarrows$	$2\,Cl^-$	2087.08
Cobalt	Co^{2+}	$\rightleftarrows$	Co	2256.42
Rhenium	Re^{3+}	$\rightleftarrows$	$Re(s)$	2722.89

(*Continued*)

Table D.1 (*Continued*)

Element	Redox reaction			Volumetric energy density [Wh/L]
Iron	Fe^{3+}	$\rightleftarrows$	Fe^{2+}	2909.77
Tellurium	$Te(s)$	$\rightleftarrows$	Te^{2-}	2992.34
Sodium	Na^+	$\rightleftarrows$	Na	3058.18
Mercury	Hg^{2+}	$\rightleftarrows$	Hg	3239.04
Gallium	Ga^{3+}	$\rightleftarrows$	Ga	3610.32
Radium	Ra^{2+}	$\rightleftarrows$	$Ra(s)$	3652.17
Barium	Ba^{2+}	$\rightleftarrows$	$Ba(s)$	4125.85
Tantalum	Ta^{3+}	$\rightleftarrows$	$Ta(s)$	4440.07
Zinc	Zn^{2+}	$\rightleftarrows$	Zn	4448.92
Strontium	Sr^{2+}	$\rightleftarrows$	Sr	4648.23
Europium	Eu^{2+}	$\rightleftarrows$	$Eu(s)$	5200.60
Palladium	Pd^{2+}	$\rightleftarrows$	$Pd(s)$	5525.92
Oxygen	$O_2 + 4\,H^+$	$\rightleftarrows$	$2\,H_2O$	5876.50
Calcium	Ca^{2+}	$\rightleftarrows$	Ca	6064.85
Fluorine	$F_2(g)$	$\rightleftarrows$	$2\,F^-$	6124.69
Lithium	Li^+	$\rightleftarrows$	Li	6280.04
Platinum	Pt^{2+}	$\rightleftarrows$	$Pt(s)$	7001.94
Vanadium	V^{2+}	$\rightleftarrows$	V	7264.92
Actinium	Ac^{3+}	$\rightleftarrows$	$Ac(s)$	7846.06
Niobium	Nb^{3+}	$\rightleftarrows$	Nb	8151.05
Chromium	Cr^{3+}	$\rightleftarrows$	Cr	8170.31
Lanthanum	La^{3+}	$\rightleftarrows$	$La(s)$	8496.20
Manganese	Mn^{2+}	$\rightleftarrows$	Mn	8554.32
Magnesium	Mg^{2+}	$\rightleftarrows$	$Mg(s)$	9053.60
Yttrium	Y^{3+}	$\rightleftarrows$	Y	9586.77
Terbium	Tb^{3+}	$\rightleftarrows$	Tb	9616.83
Titanium	Ti^{3+}	$\rightleftarrows$	$Ti(s)$	10355.60
Uranium	U^{3+}	$\rightleftarrows$	$U(s)$	10625.86
Iridium	$Ir^{3+} + 2\,H_2O$	$\rightleftarrows$	Ir	10952.32
Gold	Au^{3+}	$\rightleftarrows$	$Au(s)$	11987.52
Aluminium	Al_3^+	$\rightleftarrows$	$Al(s)$	13356.00
Zirconium	$ZrO(OH)_2 + H_2O$	$\rightleftarrows$	$Zr + 4\,OH^-$	18044.06
Beryllium	Be_2^+	$\rightleftarrows$	$Be(s)$	20334.34

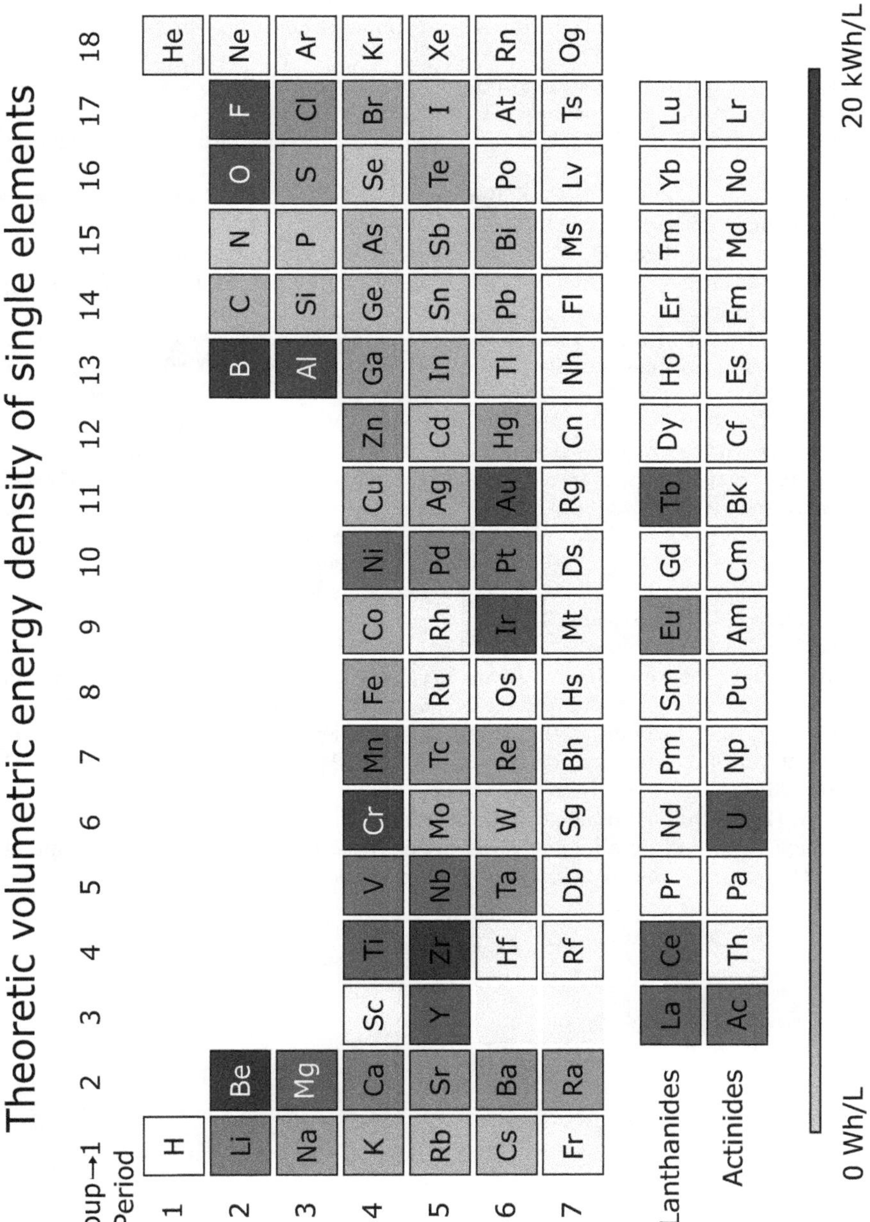

Fig. D.1 Volumetric energy density of each element in the periodic table.

Bibliography

[1] "Electrochemical series," *Internetchemie.info*, http://www.internetchemie.info/chemie-lexikon/daten/e/elektrochemische-spannungsreihe.php (in German). Date accessed: 13 September 2017.

[2] "Periodic system," *Periodensystem.info*, http://www.periodensystem.info/perioden system (in German). Date accessed: 13 September 2017.

[3] Linden, D. and Reddy, T. B. (2002). *Handbook of Batteries*, 3rd Edition (McGraw-Hill).

[4] (eds.) Imanishi, N., Luntz, A. C., and Bruce, P. G. (2014). *The Lithium Air Battery: Fundamentals* (Springer).

[5] Radin, M. D. and Sigel, D. J. (2015). Non-aqueous metal-oxygen batteries: Past, present, and future, in *Rechargeable Batteries, Green Energy and Technology*, Zhang, Z. and Zhang, S. S. (eds.).

[6] Graphene oxide for Lithium-Sulfur batteries, *IDTechEx*, https://www.idtechex.com/research/articles/graphene-oxide-for-lithium-sulfur-batteries-00008050.asp. Date accessed: 29 September 2017.

[7] Tian, H., Xia, F., Wang, X., He, W., and Han, W. (2015). High capacity group-IV elements (Si, Ge, Sn) based anodes for lithium-ion batteries, *J. Materiomics* **1**, pp. 153-169, http://www.sciencedirect.com/science/article/pii/S2352847815000477. Date accessed: 19 October 2017.

[8] Kam, K. C. and Doeff, M. M. (2012) Electrode materials for lithium-ion batteries, *Material Matters* **7**, 1, pp. 56–60, https://www.sigmaaldrich.com/technical-documents/articles/material-matters/electrode-materials-for-lithium-ion-batteries.html. Date accessed: 19 October 2017.

[9] Imanishi, N. and Yamamoto, O. (2014) Rechargeable lithium — Air batteries: Characteristics and prospects, *Mater. Today* **17**, 1, pp. 24–30, http://www.sciencedirect.com/science/article/pii/S1369702113004586. Date accessed: 26 October 2017.

[10] Zhong, Y. (2011) Lithium-air batteries: An overview, Coursework, Stanford University, http://large.stanford.edu/courses/2011/ph240/zhong2/. Date accessed: 25 October 2017.

[11] Chen, A. (2013) Sulfur-graphene oxide material for lithium-sulfur battery cathodes, *Berkeley Lab*, https://eta.lbl.gov/news/article/56320/sulfur-graphene-oxide-material-for-lithium-sulfur-battery-cathodes. Date accessed: 25 October 2017.

Chapter 2

Lithium-ion Cells: Discussion of Different Cell Housings

Kai Peter Birke and Shkendije Demolli

Currently there exist three different cell types — cylindrical cell, prismatic cell and pouch cell — that can be found in different applications, such as mobile applications (laptops, cell phones, etc.) or in the automotive sector. In the automotive sector, lithium-ion cells are becoming increasingly important. As a result, the best housing case for these cells in HEV-, PHEV-, and EV-applications is dependent on their shapes and material characteristics, even though it should be assumed that there is more than one solution for all applications. However, the best type of cell will dominate, combining a simple cell production with an effective thermal management and durability during its lifetime.

2.1 Cell Housings

The housing protects the cells from external influences and forms a mechanical integrity. The main differences of the cell housings are shape-related. For example, rigid cell housings made of steel or aluminium are used for cylindrical or prismatic cells, whereas for pouch cells the electrochemical active layer is packed in flexible aluminium laminate. Pouch cells are therefore also called "Coffee bag" or "Soft pack", as illustrated in Fig. 2.1. The cylindrical cell housing has a round cross section and the prismatic cell housing has a rectangular cross section. An overview of the benefits and detriments of the three cell housings is given in Table 2.1.

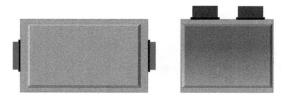

Fig. 2.1   Pouch cell ("Coffee bag"/"Soft pack").

2.2 Cylindrical Cells

The most common design of cylindrical cells is the 18650 (diameter 18 mm, height 65 mm). The electrochemical active layers are coiled symmetrically and are enclosed by the cylindrical housing made of steel or aluminium. This cell is used in notebooks, electric power tools as well as in electric vehicles [Ayub (2017)]. For the latter, several thousand cells per battery are stacked together.

Many years of experience and ongoing production improvements of high quantities lead to high quality and economical manufacturing. Other advantages are high specific energy and safety, due to the good mechanical stability [Stevens (2016); Battery University (2011)]. However, there are limiting factors in thermal management, caused by its compact design, unfavorable core-to-surface distance and small outer surface. This poses a huge challenge to cooling systems for cylindric cells with high performance and/or high capacity.

2.3 Prismatic Cells

Due to the rectangular base, prismatic cells and pouch cells enhance space utilization and can be stacked quite easily. Though their flexible design slightly increases the manufacturing costs, they can be packed more densely into a battery pack as cylindrical cells. Those cells exist either as an elliptical jelly roll or as stacked layers. The electrodes of individual cells can be stacked and are connected in parallel or consist of elliptical coils [Battery University (2011); Epec (2017)]. The big difference between prismatic cells and pouch cells is their housing type. Whereas prismatic cells are encased in a rigid metal housing as the cylindrical cell, the housing design of pouch cells encapsulates the cell layers. The flexible outer layer of pouch cells resembles shiny metallic vacuum packs of ground coffee. That is why they are also called "Coffee bag".

Table 2.1 Comparison of hardcase cells (cylindrical cell, prismatic cell) and pouch cell; 0: Weakest, 3: Best (based on [Ayub (2017); Birke (2014)]).

Criteria	Round cell (Cylindrical cell)	Prismatic cell	Pouch cell (Softpack, "Coffee bag", cell in aluminium laminated foil)	Comments
1. Volumetric energy density (Wh/l), Cell level	3	2	2	The conductors of the pouch cell stand out and are calculated as part of the volume.
2. Volumetric energy density, System level	1.5	2	2	With the round cell, only cubic close sphere packing is possible. The conductor of the pouch cell can be cranked in the system. With prismatic cells, there is the possibility of an extremely simple and compact stacking.
3. Gravimetric energy density (Wh/kg), Cell level	3	2	3	Benefits increase with the size of the pouch cell.
4. Gravimetric energy density (Wh/kg), System level *18650	2.5*	2	1.5	Still high losses in all systems. Currently, the pouch cell cannot yet use its advantages to achieve a better grade in this discipline. Nevertheless it has the highest development potential.
5. Robustness cell housing	3	3	2	In addition to the seal the higher sensitivity of aluminium laminated foil is also taken into account.
6. Flexibility format and manufacturing	2	2.5	3	The pouch cell scores when it comes to formats, cell families are easily possible through variations in thickness. However, currently the cell is rigid and not a triangle, hexagon or a flexible foil.
7. Production speed	3*	2.5	2.5	Without any additional inner conductor, round coil production is fastest. In case of additional inner conductors this advantage is lost.

(Continued)

Table 2.1 (*Continued*)

Criteria	Round cell (Cylindrical cell)	Prismatic cell	Pouch cell (Softpack, "Coffee bag", cell in aluminium laminated foil)	Comments
8. Connection between the inner to the outer cell conductor (in case of many inner conductor)/ Implementation on pole cross section	1.5	2.5	3	This is easiest to solve for the pouch cell because of its construction.
9. System level (Cell integration)	2	2.5	2	At the moment the pouch cell is the biggest challenge here. Prismatic cell does not reach 3 due to the effort of bracing (forces and precision).
10. Thermal benefits	1.5	2	3	Because of its geometry, the round cell is usually difficult to cool. The prismatic cell has a good surface, but contact of the electrodes to the housing cannot be compared to the pouch cell (vacuum). Pouch cells allow for flat conductors on both sides of the cell, so good, thermal distribution and no heat build-up at conductors.
11. Contact pressure of electrodes possible through external tension	0.5	0.5	3	Not possible for cells with rigid housing.
12. Vibration	1.5	2	2.5	The flexible outer skin makes the pouch cell less vulnerable to vibration.
13. Number of parts with cell housing	1.5	1.5	3	Due to the way it is constructed, the pouch cell has the smallest number of parts.
Total evaluation	2.0	2.1	2.5	

Nonetheless, a stable and strong housing case is needed for automotive applications, which have a long battery life of more than 10 years. Thus, the challenge is to find a flexible but sufficiently strong metal to replace the "solid can". Using a mere metal coating could lead to micro-defects, the so-called pinholes, and is therefore insufficient. A coated metal foil with a solid metal core is a better alternative for mechanical reasons. To fulfill the requirements a 40 μm thick aluminium metal foil can be used. This foil is an effective barrier with an almost unlimited lifetime [Epec (2017)]. Figure 2.2 compares the prismatic cell and pouch cell.

2.4 Stabilization of Electrode and Separator Layers

There are various advantages of using this type of foil, especially for the interior of the cell. It consists of several layers: Electrodes and separator impregnated with an electrolyte. Vacuum packaging with the foil automatically adjusts the foil and cell interior and fixes these layers. The design of the housing is greatly simplified, thus significantly fewer components are required for the cell housing. One major benefit is the flexible housing design, as the internal electrodes can be compressed directly by external pressure. The pressure is managed and distributed by means of intermediate tapes between the cells. This ensures that the separator and electrode layers are held together at high temperatures (from 60°C) or at overloads, even if gas is produced by side reactions. The risk that the formed gas could separate the layers no longer exists, indeed it increases the operational lifetime of the cell. When mechanical stress occurs, i.e., penetration by a steel nail, external pressure prevents the separator from shrinking in the presence of the compressed electrode and separator layers. Furthermore, it is advantageous that a displacement of the electrode relative to the separator is avoided, because it will prevent the possibility of short circuits. High volume changes in charge/discharge can be counteracted by new materials such as SiC composites on the negative side and NiCoMn composites with high Ni content on the positive side.

A compression of the rigid housing is not possible without deforming the cell and causing damage to it. During vibration, the current collectors collide against or tear off the cell's top, hence that is why there is an inherent risk of mechanical damage to its rigid housing, see Fig. 2.3. Another disadvantage is that the separation layer can be damaged during charging, due to the expansion of the cell coil. This may lead to the cell being strongly compressed against the housing. This is avoided in the pouch cell because

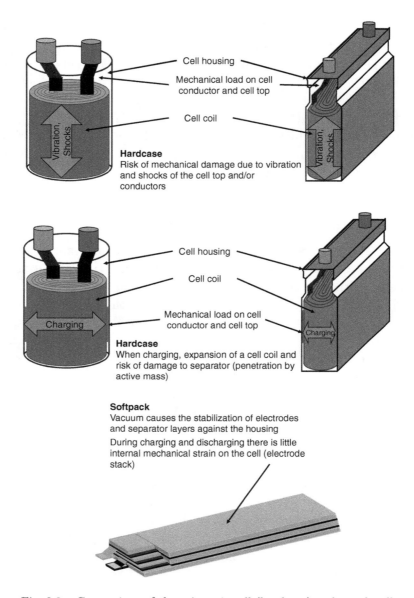

Fig. 2.2 Comparison of the prismatic cell (hardcase) and pouch cell.

of its vacuum. This stablizes the electrode and separator layers against its housing case. However, despite its sufficient flexibility, even the pouch cell will still experience very low mechanical stresses internally (electrode stack) during the charging/discharging. Thus, the design of the pouch cell, i.e., the vacuum-packed foil, is supposed to protect it from external and internal

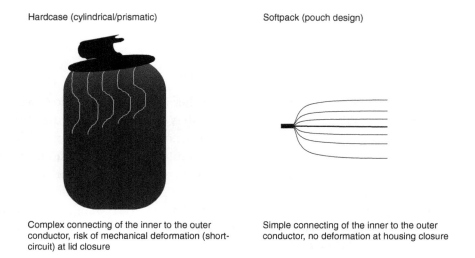

Fig. 2.3 Comparison of hardcase cell (left) and pouch cell (right) during vibration.

influences. A lid is not required; a seal at the end of the manufacturing process is sufficient.

2.5 Gas Evolution

A major safety issue in battery development is gas evolution, due to overcharging. Surpassing a maximum tolerable value in the pressure and temperature build-up gas has to be discharged via an opening mechanism. In the life cycle of the battery the opening parameters may change because of aging processes, such as structural distortion in the cell. It is an advantage if this opening mechanism is already integrated in the cell level. However, such an integration is accompanied by a higher number of components and more complex and costly production processes.

In rigid housing cells, a kind of predetermined breaking point, which is also called "bursting membrane", is installed. This membrane, which is located in a predefined location in the cell, opens only when the maximum value is exceeded, to allow the cell to vent quickly and accurately. In contrast, the pouch cell automatically opens a sealed seam at high temperatures and cell internal gas pressures. This opening mechanism is independent of any particular mechanical pressure, as compared to cells with a rigid housing. However, a disadvantage of the pouch cell is that this seal does not open at any particular location, which makes fast venting difficult. One possible solution would be to create an isolated gas space by arranging the cells appropriately [Epec (2017)].

2.6 Flexibility with Respect to Cell Size

The main advantages of the pouch cell over other cell types are its low weight and improved flexibility in production — there is no single standardized cell; each cell is unique, depending on the application and manufactuer [Epec (2017)]. This scenario is made possible by a simple deep drawing and cutting of the aluminum laminate, since the electrodes and separators for lithium-ion cells can be easily cut to size. Other cylindrical and prismatic cells are prepared by a standardized method [Battery University (2011)]. Each time the case design changes, the production set-up has to be modified, which increases production time and cost, not to mention affecting product quality.

Aluminum is used because its plasticity can be used as a material for the metal core of the outer foil. The aluminum laminate is composed of three layers — a 40 μm aluminum layer sandwiched between two layers of plastic materials. Mechanical protection is provided by a 30 to 35 μm thick outer layer, with the 80 μm thick inner layer acting as heat lamination of the housing case at approximately 200°C. During lamination, the thickness of the aluminum laminate is reduced by about 30 to 40 μm. By this process, the single foil layer within the unsealed area is 150 to 160 μm thick, whereas the sealed seam is 270 to 280 μm thick.

However, it is proprietary knowledge among Al-laminate suppliers on how the inner plastic layer is made. To protect the aluminum foil from hydrofluoric acid which is found in traces in the electrolyte, the inner sealing layer must contain a blocker. The hydrofluoric acid is not concentrated enough to be harmful, but it could damage the aluminum layer.

2.7 Producing Pouch Cells

Pouch cells are not only lightweight, but can be designed to be very slim. This is advantageous because many standard, and above all, new formats can be produced and customized for specific applications [Battery University (2011)]. Nonetheless, there are challenges with large cell areas and thicker cells. For example, thicknesses above 15 mm could result in the foil tearing during deep drawing. Furthermore, the ratio of sealed seam to body (area of the active electrodes) with thicker cells (>10 mm) and low available construction height results in a noticeable impairment of the pouch cell. Here prismatic hardcase cells have a significant advantage over the pouch cells.

Another important component of the cells is the prevention of water ingress and the loss of electrolyte through evaporation. The metal layer of the outer cover of the pouch cell guarantees this reliably. Though, this is not the case for the plastic areas of the 280 μm thick sealing seam. Thus, the volatile solvents of the electrolyte can escape and water can still penetrate. This could lead to a very small additional capacity loss and an associated increase in internal resistance. Indeed, measurements and experience over many years show that the effects of other cell-internal electrochemical aging processes — over the lifetime of the cell — outweigh the effects of possible water ingress and electrolyte leakage. The quality of the sealed seam can be accurately checked during the cell production process for example, by ultrasonic quality control. This ensures that there are no traces of electrolyte, inclusions or faulty seals.

2.8 Status Quo of Cell Concepts

The current and most important properties of the cells are evaluated and summarized in Fig. 2.4. It should be noted that the application and the overall system is decisive to the choice of cell architecture. With regard to increasing energy density and high capacity in cells, all three cell formats will reach their limits. For example, the security challenges for these areas are difficult or cannot be fulfilled. For automotive applications, the current

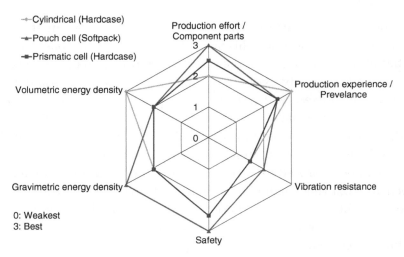

Fig. 2.4 Overview of the important properties of cylindrical, prismatic (hardcase) and pouch cells.

technical limit is restricted to approximately 100 to 120 Ah per cell and 250 to 300 Wh/kg.

The cylindrical cell 18650 was developed for notebooks and electrical tools, but it is also used in the automotive sector [Ayub (2017)]. Because of its extremely high-energy density of up to 250 Wh/kg at the cell level, as well as its high and fast mass production, it has established itself in the electric vehicle market. However, there are several disadvantages, especially when large cells are needed. For starters, the configuration of large cylindrical cells (about 50 Ah) is a very big challenge, which is why they are not used. Furthermore, it is difficult to arrange both poles on one side and a multiple contact is required for the current flow. Compared to prismatic cells, cylindrical cells are much easier to produce due to their low capacity per cell. However, this production benefit is considerably reduced because this process has to be interrupted every few turns in order to weld the layers for multiple contacting. As a result, the only option is to use the 18650 cylindrical cell for large EV batteries, combining several 1000 cells. In the high voltage range (HEV, PHEV and EV), cylindrical cells will therefore increasingly be replaced by prismatic or pouch cells. With these cells the 50 Ah is already achieved, which is why mass production has to become economical by achieving volumes corresponding to the 18650 cells [Messina (2015)].

2.9 Outlook

For economical use of battery cells, the production of a few but standardized cells plays an important role. This not only has a significant positive effect on costs, but also on the quality of the cells. Standardization at the cellular level is possible and it already exists in a few cell types. Module standardization has many advantages, which is why preliminary experiments are being made in this field. Concatenating the battery cell to installation spaces and other components (electromechanical parts) increases the exploitation of the development potential from the system level. Thus, the energy density (Wh/kg) can be improved from cell level to system level. In particular, pouch cells benefit from their high-energy density (Wh/kg) at the cell level. Being able to shift these at the system level, and thus avoiding minimization by other components, would be a big step forward.

While the cell mainly contributes to performance qualities such as energy and power density, durability, safety and cost, a strong integration of electric mobility with road traffic is dependent on the available system and module installation spaces. The pouch cells, when compared to other cell types, are favored for use in mobility because of these factors: its possibility to directly compress electrodes at cell level, thermal benefits and simple housing production. Finally, because of the above-mentioned cells, those who are already looking into the development of solutions with regard to cell and system interaction, will achieve a significant leading edge in the mass production of electric vehicles.

Bibliography

Ayub, I. (2017). Introduction to lithium-ion rechargeable battery design, http://www.edn.com/design/power-management/4458054/Introduction-to-lithium-ion-rechargeable-battery-design.

Stevens, M. (2015). Lithium ion battery energy storage systems using 18650 cells, https://evgrid.com/lithium-ion-battery-energy-storagesystems-using-18650-cells/.

Battery University (2011). Types of battery cells, http://www.batteryuniversity.com/learn/article/types_of_battery_cells.

Epec (2017). Prismatic and pouch battery packs, http://www.epectec.com/batteries/prismatic-pouch-packs.html.

Birke, P. (2014). Lithium-ion accumulators.

Messina, C. (2015). Lithium-ion cylindrical cells vs. prismatic cells, http://www.relionbattery.com/blog/lithium-cells-should-i-go-cylindrica.

Chapter 3

Integral Battery Architecture with Cylindrical Cells as Structural Elements

Christoph Bolsinger, Marcel Berner and Kai Peter Birke

Battery architecture for electromobility is driven by continuous energy density enhancement. This counts for weight and volume as well as costs. Nowadays, much effort is put into the cell level improvements. However, the situation on the battery level is surprisingly lacking in focus. The additional mass caused by the battery/module housing, safety devices and other peripheral components lower the relative proportion of the active electrode materials on the total weight. In current battery systems the reduction of the specific energy and power density can cause 50% of the energy density on cell level. Figure 3.1 depicts the loss of specific energy density from cell to system level. Additionally, those batteries show a complex recycling process in case of very rigid housing and welded cell connectors.

    The enormous potential of energy density enhancement on the system level requires new construction methods, since the energy density enhancement of lithium-based cells will be more and more limited within the next decade. The Project "LIBELLE", founded by the VECTOR STIFTUNG shows new approaches where the cell is used as a self-supporting part of the battery system. One cylindrical 18650 battery cell can withstand an axial statistic load test of up to several kilo newtons. Therefore, a multiple array easily sustains automotive standard test conditions. The final battery housing has thus no tasks to protect the battery against mechanical impacts and can therefore be constructed with very lightweight materials. The cooling

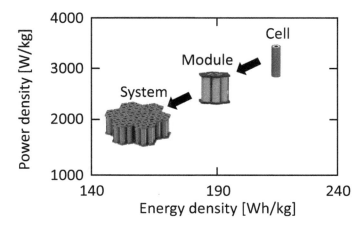

Fig. 3.1 Loss of energy and power density from cell level to system level due to additional housing and peripheral components. The weight from the cell to the module level increases in case of the cell connectors, cell holders and cell supervision unit. The housing, contactors, wires, fuses, different sensors and thermal management further increase the weight from module to system level. Values according to Ketterer *et al.* (2009).

plate will be made of modern plastic. Additionally, a recycling concept is implemented at the cell level. The modules can be completely disassembled since the cells are screwed on and clamped. The recycling concept is completed with a cell supervision unit (CSC) which communicates over an optical communication channel.

Generally, battery production concepts still show a huge lack of industrialization, which results in an cost-intensive process. The ratio between the cell and battery (without cells) costs should be about 80:20, or even more for the cell, to achieve competitive costs compared to combustion engines. Current batteries show a ratio of about 40 (or even less):60. This makes the battery system the cost driver of current electric vehicles (EV). Therefore, the LIBELLE-project provides a simple assembly concept where the cell becomes a construction element. This may help to achieve cost reductions. Additionally, a modular expandable concept offers the application from electric scooters to large electric vehicles. The performance of the battery can be adjusted by employing high-power or high-energy cells. The heat conduction is carried over the ground and the bottom plate where the cells are mounted. These plates are cooled by an plastic cooling plate.

The following sections describe state of the art battery systems and their design considerations, as well as new approaches and their implementation.

3.1 State of the Art Battery Systems

A battery module consists of a serial connection of single cells. The number of cells defines the required voltage range of the module. Depending on the application, several modules or single cells are connected in parallel to increase the capacity. The connected battery cells and optional periphery, such as the cell supervision circuit (CSC), battery management system (BMS), thermal management and contactors, build the battery system. The battery system's design considers the specific characteristics of the utilized battery cells to ensure their reliability, safety, efficiency and ideal mode of operation over their life span. The design of the battery modules can be separated into two architectures — block and modular. Both are described in Secs. 3.1.1 and 3.1.2. A further design aspect is the cell connection (circuitry) in the modules which is described in Sec. 3.1.3.

3.1.1 *Block architecture*

The battery system is either designed as a modular or block architecture. Figure 3.2 shows a battery with a block architecture. All battery cells as well as all peripheral components are inside one battery housing. On one hand this reduces space because all cells are directly connected with each other and there is no separation. On the other hand the complexity level rises during the assembling and dissembling processes, potentially because of the

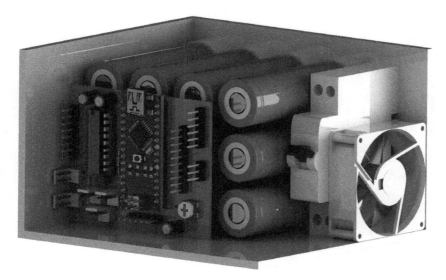

Fig. 3.2 Battery in a block architecture. Instead of building modules, all battery cells are directly connected and packed with all peripheral components in one battery housing.

high voltages of the battery and it increases the risk of a thermal runaway propagation from one cell to the surrounding cells.

3.1.2 Modular architecture

Individual modules, consisting of an appropriate number of single cells, are interconnected to built the battery. Additionally, each module contains specific peripherals like the CSC, contactors and cooling interfaces. This simplifies the handling during the assembling process in case of smaller module voltages. Additionally, the recycling and maintenance process is simplified because of smaller exchangeable modules. Furthermore, the additional module housing limits the risk of thermal propagation from one module to another. Figure 3.3 shows a battery in a modular architecture. The arrangement of the different modules depends on the shape of the battery and their location in the vehicle. In this case the battery is located between the front and the rear axis. Generally, large battery systems use a modular architecture in case of high variability and better handling of low module voltages.

3.1.3 Cell circuitry

Automotive applications require high voltages to reduce the ohmic loss of the power lines. The majority of pure electric vehicles use battery voltages of about 400 V. In future batteries voltages will increase to about 800 V. The higher the voltages, the lower the ohmic losses but the higher

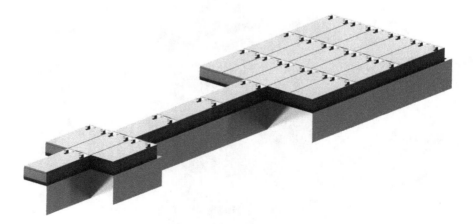

Fig. 3.3 A battery in a modular architecture. Different modules are connected together to build the battery. Each module contains a cell supervision circuit which monitors the operation of the single cells. A battery tray holds the modules and the peripheral components. A cover protects the battery against electromagnetic radiation as well as dust and dirt.

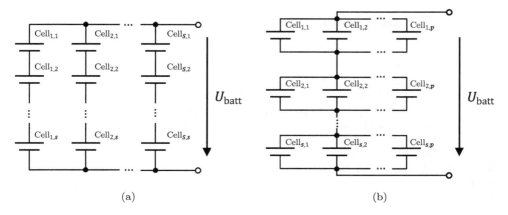

Fig. 3.4 Different possible cell circuitries to enhance the battery voltage U_{batt} and capacitance. (a) shows the parallel connection of serial-connected cells (strings) and (b) shows the serial connection of parallel-connected cells. For equal cell voltages the battery voltage U_{batt} for (a) and (b) shows the same value. For (a) all cell voltages have to be monitored. The parallel battery cells in (b) balance themselves. Therefore, only one voltage of the parallel connection has to be monitored.

the requirements on the system because of the aspects of high voltage engineering. Battery modules with high voltages and capacities consist of a number of single cells that can be interconnected in different ways. Figure 3.4(a) shows the parallel connection of strings. One string consists of the serial connection of cells. The CSC monitors each cell in the strings to ensure the operating conditions. This means a high wiring effort and an additional complexity. A further disadvantage is the lack of redundancy. If any cell falls out, the whole string is inoperable. Figure 3.4(b) depicts the serial connection of parallel-connected cells which increase the redundancy and lowers the wiring effort. Due to the self balancing process of parallel-connected cells, the CSC only monitors one voltage per serial-connected module (parallel-connected cells). If one cell is cut off, the string still remains operable.

3.2 The Battery Cell as a Structural Element

A crucial design criterion for battery systems is the determination of an appropriate battery cell. Figure 3.5(a) to (e) shows the three general lithium-ion cell formates. The electrodes can either be housed in a cylindrical or prismatic rigid case or in a flexible "pouch bag". The so-called "pouch bag" consists of a plastic-coated aluminum foil. The following subsections describe the properties of each format regarding the design of a battery module.

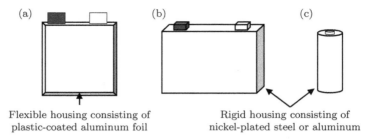

Flexible housing consisting of plastic-coated aluminum foil

Rigid housing consisting of nickel-plated steel or aluminum

Fig. 3.5 The different types of lithium-ion cells. (a) shows the prismatic geometry with a flexible pouch bag as a housing. (b) and (c) depict the prismatic and cylindrical cell types with a rigid housing made of nickel plated steel or aluminum.

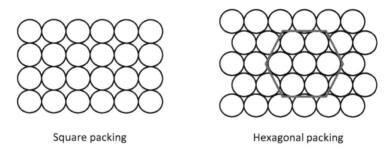

Square packing

Hexagonal packing

Fig. 3.6 Square and hexagonal packaging principle for cylindrical battery cells. Seven cells arranged in a hexagonal lattice in a honeycomb provide the maximum packing density of 90%. Hence, the submodule uses honeycomb geometry.

3.2.1 Cylindrical cells

The cylindrical cell is well known in the consumer sector. The most common type is the 18650 battery cell with a 18 mm diameter and 65 mm height. Also, electric vehicles use thousands of this type in their large batteries (Tesla). The electrodes are wounded and than stacked into a steel or aluminum cylinder. The advantages of the cylindrical battery cell are the compact design and the highly experienced production process, which mean large numbers of units and therefore low costs. On the other hand, the compact design and the low ratio between cell core (volume) and cell surface causes a low cooling efficiency. Especially for high-power or high-energy cells this effect limits the operation. Figure 3.6 shows the hexagonal lattice and the square lattice which provides two options of packing the cylindrical cells in the battery modules. On the battery system level the cylindrical geometry limits the packing density to a theoretical maximum of around 90% for a hexagonal honeycomb-shaped packing. In practice, the safety distances, the cell mounting and the space for cooling units can reduce the packing density to values of about 60%.

3.2.2 Prismatic cells

The solid case of the prismatic battery cells provides an optimal geometry for packing without any cavities. The electrodes are either implemented as an elliptic coil or as a stack. The prismatic shape of the cell offers a high surface to volume ratio and therefore good thermal properties. Even better thermal properties show the prismatic pouch cells in case of their elastic thin aluminum case and large area tabs. Cooling the tabs leads the heat out of the cell in a very efficient way. However, the voltage difference between the battery tabs has to be considered in order to prevent short circuits over the cooling parts.

3.2.3 Battery cells as structural elements

The rigid case of the cylindrical and the prismatic cell resists shocks and static forces and therefore protects the electrodes. Figure 3.7 shows the force displacement curves of three cylindrical 18650 battery cells with axial

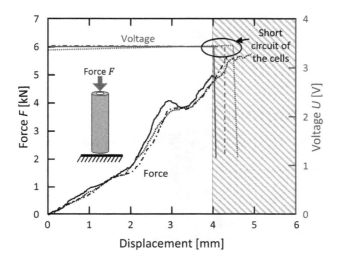

Fig. 3.7 Stress strain curves of three axial-loaded 18650 cylindrical battery cells with monitored voltages. Up to a force of about 1.5 kN, the curves show a linear behavior, which describes an elastic behavior. At axial forces of ~5 kN the monitored voltages break down in case of internal shortcuts of the separator. The concept of the battery module with integral single cells utilizes the area of elastic deformation. Values according to Zhu et al. (2016).

compression [Zhu et al. (2016)]. The cells withstand several kilo newtons (kN) until the deformation causes an internal short circuit. This happens at a force of about 5 kN and can be identified on the voltage drop. For a force of about 1.5 kN the displacement increases proportionally with the force which represents an elastic deformation. In state of the art battery systems, an additional battery housing protects the cells against mechanical shocks and deformation.

The new approach for the battery architecture presented in the following utilizes the rigid cell housing as a structural element of the battery system. Therefore, the area of the elastic deformation is used. This approach reduces weight in case of the battery housing being reduced. In this case the battery housing is designed only to seal the cells against moisture and pollutant. An appropriate arrangement of the cells ensures an even distribution of the mechanical impacts. Figure 3.8 shows the principle of an integral battery architecture. The same approach was used in the automotive industry to reduce weight and cost when the unitized body structure was introduced in the 1920s. Before then, passenger cars consisted of a rigid frame or chassis, which carried the separate body and other components, as the engine and the drivetrain. In contrast the unitized body was a one-piece structure which integrated the chassis and bodywork of the car. Therefore all loads were carried by the car itself. In general the unitized body structure is more rigid than the body-on-frame structure. This work transfers the process to the battery systems design.

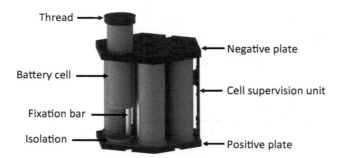

Fig. 3.8 Construction of the submodule for a modular battery architecture with integral single cells. A thread is pressed on the negative pole of each battery cell. The cell is then put through the negative plate. With the tightening of the thread into the negative plate, the positive pole of the cell is clamped on to the positive plate. This connects the positive and the negative plate electrically, mechanically and thermally with the corresponding terminal of the battery cell. The fixation bars only counteract the generated tensile force.

3.3 Construction of the Battery Module

The following section describes the boundary conditions and the requirements on a battery system with integral single cells, and their implementation in the design process. The main issue is the utilization of the rigid cell housing as a part of the integral battery system. This reduces weight and raises the gravimetric energy density on the system level. Furthermore, the design process considers two additional aspects — a modular architecture which makes the battery suitable for as many as possible applications (large limousines, small city cars, scooter, e-bikes, etc.) and an easy concept to replace single battery cells, which reduces the effort of recycling processes and improves maintenance.

The design of the battery module must consider the following requirements:

- The contact resistance of the cell connections has to be lower than a hundredth of the DC-resistance of the cell.
- Moisture-proof to avoid contact corrosion.
- A life cycle of higher than ten years.
- Automotive standards.
- Construction with integral single cells and no further load-bearing elements.
- Integration of thermal management.
- Safety aspects (overload, deep discharge, overcurrent, temperature, etc.).
- Module scalability for variable capacity and voltages.
- Easy concept to replace single battery cells.
- Maximum weight increase of factor 1.3 from cell to system level.
- Providing gas channels in case of battery cell venting.

Figure 3.8 shows the submodule with seven cylindrical cells. A thread is pressed on the blank can (negative pole) and threaded into the negative plate of the submodule. The tightening of the thread clamps the positive pole to the positive plate of the submodule. Thus, the battery poles are electrically, mechanically and thermally connected to their respective plate. Plastic rings isolate the can against the positive plate and thus prevents an electrical short circuit. Fixation bars counteract the tensile force generated by the preload of screwed cells.

3.3.1 *Cell connection*

The contact resistance which occurs between the cell terminals and the cell connectors should preferably have lower than a hundredth of the

DC-resistance of the cell. With this rule of thumb, the power loss due to the cell connection is negligible in relation to the power loss of the battery cell. A high contact resistance leads to a significant heat generation which is transferred into the battery cell, causing faster degradation and a higher risk factor. Additionally, it lowers the efficiency and power capability. A nonseparable joint between the battery cell and the connector via welding, soldering or gluing processes is not appropriate in case of the requirement of an easy concept to replace single battery cells. Therefore, clamped joints are used to interconnect the battery cells. Section 3.5 describes the application and characteristics of clamped battery cell connectors in detail.

3.3.2 *Moisture proof*

It is important for the battery to be moisture-proof so as to protect the components, and especially the conductors, against corrosion and wear. Contact corrosion occurs between two contact elements (connected metals) with different electrode potentials in an electrolyte (water, moisture, etc.). Being moisture-proof can only be realized by a surrounding housing around the battery that will protect it against moisture and pollution. It does not protect the battery against mechanical impacts (see Sec. 3.2). Therefore, it is also possible to use a composite-layer film similar to a lithium-ion cell's pouch bag or lightweight plastic materials.

3.3.3 *Lifetime*

A **lifetime** of more than ten years is a benchmark for a passenger car and is therefore also required for the battery of an electrical vehicle. A crucial aspect to achieve this maximum lifetime is to adhere to the automotive standards and pass all respective tests, which include mechanical shock, vibration and drop tests as well as electrical and thermal tests. In addition to the operating temperature, a homogeneous temperature distribution within the battery and the cells ensures an extended lifetime [Hunt *et al.* (2016); Klett *et al.* (2014)]. Section 3.6 describes the thermal management in detail.

3.3.4 *Automotive standards*

The release of a battery pack into the market requires the **compliance to automotive standards**. Table 3.1 lists the most important international standards and regulations for the safety testing of lithium-ion batteries, modules and cells for automotive applications [Ruiz *et al.* (2017)]. These standards include specifications of environmental, electrical and mechanical

Table 3.1 List of the most important standards for the safety testing of battery packs, modules and cells in automotive applications, according to Ruiz et al. (2017).

Standard	Description
SAE J2464	Electric and hybrid electric vehicle rechargeable energy storage system safety and abuse testing
SAE J2929	Safety standards for electric and hybrid vehicle propulsion battery systems utilizing lithium-based rechargeable cells
ISO 12405-1	Test specification for lithium-ion traction battery packs and systems, Part 1: High-power applications;
ISO 12405-2	Test specification for lithium-ion traction battery packs and systems, Part 2: High-energy applications
ISO 12405-3	Test specification for lithium-ion traction battery packs and systems, Part 3: Safety performance requirements
IEC 62660-2	Standards for secondary lithium-ion cells for the propulsion of EVs, Part 2: Reliability and abuse testing
IEC 62660-3	Rechargeable cells standards publication secondary lithium-ion cells for the propulsion of electric road vehicles, Part 3: Safety requirements of cells and modules
UN/ECE-R100.02	Uniform provisions concerning the approval of vehicles with regard to specific requirements for the electric power train
UL 2580	Batteries for use in electric vehicles
USABC	United States advanced battery consortium — electrochemical storage system abuse test procedure manual
FreedomCAR	Electrical energy storage systems abuse test manual for EV and HEV applications
KMVSS 18-3	Korea Motor Vehicle Safety Standards for traction batteries
AIS-048	Battery operated vehicles — safety requirements of traction batteries
QC/T 743	Lithium-ion batteries for electric vehicles. Chinese voluntary standards for automobiles

abuse tests. These test procedures have to be considered in the early design process. Further standards describe the transportation and storage of battery cells. Additionally, manufacturers' instructions also have to be considered.

3.3.5 No further load bearing elements

The use of nonstructural elements (except the single battery cells) than the single cells is one of the main issues for battery module design to reduce weight and costs. Figure 3.9 shows the frame of the submodule with nonstructural elements and without cells. This module uses cylindrical battery cells in the form of the 18650 geometry because of its housing's high structural stability and its high gravimetric energy density. The

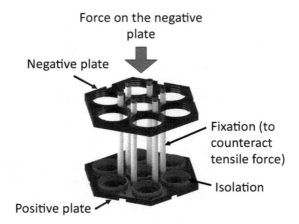

Fig. 3.9 Frame of the submodule without battery cells. The fixation bars only counteract the preload of the screwed battery cells. If a force acts on the pure frame, the plates will get together.

measurements of Zhu *et al.* (2016) show that a cylindrical 18650 battery cell withstands axial stress of several kilo newtons until it breaks down and short circuits. Until about 1.5 kN the cell shows an elastic deformation. This force determines the number of cells in a module that will pass structural stability and mechanical shock tests. The purpose is the utilization of the elastic range within all strain tests to avoid any damage to the battery cells.

3.3.6 *Thermal management*

Thermal management extends the lifetime and performance of the battery. Therefore, module design should incorporate the ability to easily integrate a cooling system. Section 3.6 describes different cooling systems in detail. A forced convection is possible by a generated airflow between the battery cells. Air, as well as liquid, could be used. In addition, adapted cooling plates placed at the bottom and at the top of the module cool the cells via the cell terminals.

3.3.7 *Safety aspects*

A cell supervision circuit (CSC) monitors the voltage and temperature of each battery module to guarantee ideal operating conditions and to prevent safety critical states like overcharge and deep-discharge. Thus the operation of the cells in the given temperature range is assured. Section 3.4 describes the principle and a detailed function of the CSC.

3.3.8 Scalability

To achieve cost benefits the module has to be suitable for as many as possible applications. Therefore, the **scalability** with respect to electrical variables like capacity and voltage, as well as geometrical variables like the width and the height of the battery has to be realized. Conductive and isolating joints connect the submodules to larger units. This realizes parallel and serial connections. The hexagonal geometry of the submodule makes the module's width and length expendable. Even a vertical expansion is possible. Screws through the positive and the negative plates strengthen the module and provide an additional electrical contact. In this way, small battery packs are actualized for application in e-bikes or pedelecs. For application in EVs the submodules are expanded horizontally to build large battery packs that can be placed in the vehicle's underbody.

3.3.9 Exchangeable single battery cells

An **easy concept to replace single battery cells** is one of the key features of the module. This simplifies the maintenance, assembly and recycling process. A separable joint ensures the simple replaceability of the battery cells. Therefore, the cell connectors are clamped onto the cell terminals. A male thread is pressed onto the can and contacts the negative pole electrically and mechanically. Thus, the battery cell can be screwed

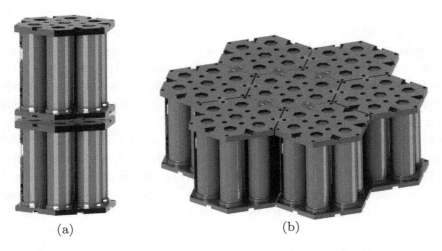

Fig. 3.10 Scalability of the submodules. The submodules can be interconnected via conductive and nonconductive module connectors. In (a) the submodules are scaled vertically, i.e., for application in an e-bike. In (b) the submodules are scaled side by side horizontally, i.e., for large EVs.

into the negative plate that clamps the positive terminal onto the positive plate. The preload of the screwed battery cell connects the thread with the negative plate. An additional isolation between the can and the positive terminal ensures the absorption of axial forces by the can and therefore protects the positive terminal against plastic deformations. An elastic or adjustable conductor like a contact spring, a conductive polymer or a grub screw determines the force of the clamped contact and therefore its contact resistance. Section 3.5 describes the contact resistance of cell connectors in detail. For example, the preload which the battery cell is tightened to corresponds to the counteracting force of the contact spring.

3.3.10 *Gas channels*

In case of an internal breakdown of the battery cell the generated pressure is released through the cell's pressure vent. For cylindrical battery cells in the 18650 format the pressure vent is usually integrated in the positive terminal. The design of the module considers gas channels that lead the generated flammable gas out of the battery to avoid an ignition or further damages. This is realized by two through-holes for each cell in the positive plate.

3.4 Integrated Cell Supervision Circuit

A battery management system assures an optimal operation of the battery cells. Monitoring of each cell voltage in a serial string is therefore one of the essential functions. An additional measurement of the temperature, e.g., at the hottest point, enables the state estimation of the cells' actual temperature. Comprehensive information about temperature distribution and temperature processes within a complex battery system is a key factor for safe and reliable operation. State of the art battery systems use a modular architecture, in which a certain number of cells are monitored by one or a few cell supervision circuits (CSC). A CSC is physically connected to the cells as well as to the higher battery management system (BMS) by wires. This results in an elaborate wiring harness that increases weight, costs and complexity. Complex wiring harnesses increases the probability of failure and limits the modularity of the systems. The battery cells supply the CSCs and the BMS. Therefore, the current consumption has to be as low as possible, especially for long idle states, i.e., when the vehicle is parked, the CSCs discharge the battery and therefore lower the range of the vehicle. Furthermore unbalanced battery cells in a serial string lowers the usable capacity of the battery as well. State of the art BMSs therefore provide balancing methods for the equalization of the cells. The balancing can either be a passive method where the energy is dissipated into heat or an active method which transfers the energy of the cells with a higher SoC to cells with a lower SoC [Daowd et al. (2011)].

The novel approach of this integrated cell supervision circuit (iCSC) is the integration of the modular printed circuit board (PCB) into the submodule. The PCB of the iCSC is clamped between the positive and the negative plates, which builds the mechanical and the electrical interface. The unnecessity of additional wiring simplifies the assembling process of the battery and enables a highly modular structure. The iCSC monitors the voltage of the single cell, measures the temperature at the cell/iCSC interface and supplies itself from the cell voltage. The iCSCs communicate their measured values to a higher BMS using an optical communication system. Hereby, a light-guiding plastic layer which is placed on the surface of the battery acts as the common transmission channel. This plastic layer is the bus medium that directly connects the iCSCs with the BMS. The bus access is organized in a master-slave principle where the iCSCs act as the slaves and the BMS as the master. Furthermore, the iCSCs provide a passive balancing method for the charge equalization of single cells and an energy saving mode which decreases the current consumption down to $9\,\mu$A. The key features of the iCSC are:

- Utilization of cost-efficient electronic components
- Optical bus communication
- Galvanic isolation between the iCSCs
- Modular structure
- Cell voltage monitoring between $0\,$V and $4.4\,$V with an accuracy of $10\,$mV
- Temperature monitoring between $-20°$C and $+80°$C
- Current consumption in an inactive state is lower than $9\,\mu$A
- Passive balancing
- Power supply of the iCSC through the mechanical interface.

3.4.1 *Balancing*

The iCSC additionally provides a balancing circuitry. If the SoC between the battery cells in a serial string differ too much from each other the iCSC of the cells with the highest SoC connects a resistor parallel to the battery cells. The resistor discharges the cells until the SoC is in the required operating range. Subsequently the iCSC disconnects the resistor from the cells. The passive balancing method can equal the SoC during the charging or

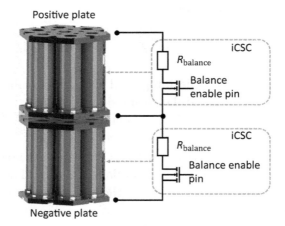

Fig. 3.11 Principle topology of the passive balancing method. The microcontroller of the iCSC can trigger a MOSFET by the balance enable pin, which connects a power resistor parallel to the submodule. This discharges the battery cells. The balancing current can be adjusted in the design process by the value of the resistor. For sufficient thermal conductivity, the thermal pad of the resistor is connected to the grounding layer of the PCB.

discharging process as well as in the idle state. Figure 3.11 shows two serial-connected submodules with the integrated iCSC and the principle balancing circuitry. The PCB leads the generated heat from the resistor to the positive and the negative plates. Since the cooling system removes the heat from the plates, high balancing currents are possible.

3.4.2 Mechanical integration

The PCB of the iCSC is mechanically integrated into the submodule during the assembling process. Therefore, the PCB is first placed between the positive and negative plates and then clamped by tightening the fixation bars. The gold-plated end caps of the PCB are pressed against the plates and thus generates a mechanical, electrical and thermal interface. This approach requires no wiring. Dissipated heat from the electronics leads the PCB through the end caps to the positive and negative plates where active cooling takes place.

3.4.3 Communication

An infrared (IR) diode at the end cap of the PCB couples the infrared radiation through drill holes in the positive or negative plate towards the

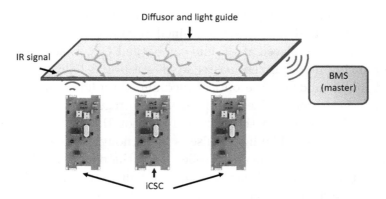

Fig. 3.12 Principle communication between the iCSCs and the higher-level BMS. Each iCSC comprises an IR-transmitter and -receiver. The IR-radiation can be coupled into the planar optical transmission channel on each position. The outer foils diffuse the light homogeneously whereas the central layer guides the light through the whole plane.

optical transmission channel. Figure 3.12 shows the communication principle between an iCSC and the BMS. The planar transmission channel consists of a stack of thin optical plastic foils. The central layer guides the light through the whole plane. The overlaying diffusion layers homogenize the light over the whole area. This cost-effective and widely available technology is well known from the backlight illumination in liquid crystal displays (LCD). The homogeneously distributed IR radiation can be received at any position on the transmission channel. An IR-receiver, which is placed next to the IR-diode on the iCSC, receives and processes the signals from the shared optical medium. The signal processing contains an electronic filter. The output of the filter is directly connected to the input of the microcontroller. A pulse-width modulation (PWM) of the data signal additionally increases the stability against distortion and ambient noise, e.g., by adjacent light sources. The chosen IR-transceiver technology is well known from remote control applications. The technology is mature, widely available as well as inexpensive. The master-slave hierarchy organizes the access on the planar optical bus medium.

3.4.4 Energy saving

The design of the iCSC considers energy saving aspects. In active state the current consumption is in the range of milliampere. However, for most of the time the iCSC might not be in use, e.g., for a parked vehicle. A current consumption in the milliampere range would discharge the battery cell significantly, and therefore, it is not negligible during the inactive phases of the idle state. The implemented power saving mode shuts down the iCSC when not used for a predetermined period of time. An additional phototransistor, also placed at the end cap of the PCB, wakes the iCSC from its power saving mode. If the master applies a bright light pulse of several tenths of milliseconds on the optical medium, the phototransistor connects the cell voltage to the enable pin of the voltage regulator of the iCSC. The cell voltage then activates the voltage regulator that starts to power the iCSC microcontroller. The light pulse is long enough to allow the iCSC to boot. After booting, the microcontroller holds the enable pin high by itself and thus stays supplied and active after the light pulse ends. If the master sends a command to shut off the slaves, the iCSCs release their enable pins and therefore shut themselves off. Furthermore, if the iCSC does not receive commands within a predetermined amount of time, an internal timer triggers the release of the enable pin. Therefore, the iCSC shuts itself off when not used for a longer time. This technology reduces the current consumption of the iCSC when it is in its sleep state to about $9\,\mu\text{A}$. The bus master can always reactivate the iCSC by a light pulse.

3.5 Cell Connectors

To achieve a simple recycling process due to exchangeable single cells, the cell connection is a crucial aspect which could be done in different ways. The following section shows the state of the art of the joining methods for battery cells and compares their characteristics. One crucial property to evaluate the quality of a joint is the electrical contact resistance (ECR), which is described in detail in Sec. 3.5.2. Section 3.5.3 describes the characteristics of clamped cell connectors which enable the simple separation of the joint.

3.5.1 *State of the art*

The connection of cylindrical battery cells is commonly done by resistance spot welding, ultrasonic welding, laser beam welding or soldering [Lee et al. (2010)]. Figure 3.13(a) to (d) shows the principal techniques for the battery cell interconnection. A huge drawback of these non-separable joints is the complex disassembling process. In contrast the clamped joints provide a simple assembling and disassembling process in which the joint takes no damage. Table 3.2 lists further benefits and drawbacks of the different technologies. Clamped joints are well established for high current applications and even for pouch cells. Therefore, the current tabs are attached to the collector bars via brackets. The submodule uses clamped

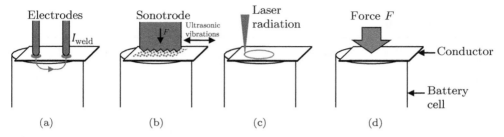

Fig. 3.13 Principal techniques for battery cell interconnection. (a) shows the resistance spot welding process whereas two electrodes are pressed on the welding partners. The current through the electrodes melts the underlaying materials in a small spot and fuses them together. The functional principle of ultrasonic welding is shown in (b). Therefore, a sonotrode which vibrates in the ultrasonic range is put on the welding partners. These are scrubbed together by the vibrations under pressure. (c) shows the principle of laser beam welding. Therefore, the joint partners are heated up by the absorption of the laser beam. In (d) the joint partners are connected by an applied force.

contacts for interconnecting the battery cells with the respective positive or negative plate. Therefore, the cell connector and the terminal of the cylindrical battery cell are pressed together by an applied contact force.

3.5.2 *Electrical contact resistance*

The electrical contact resistance (ECR) between the cell connector and the battery terminal evaluates the quality of the joint. A high ECR as well as a high bulk resistance of the cell connector leads to heat generation due to the ohmic power loss. The high thermal conductivity of the cell terminals/tabs leads the heat into the cell which causes a higher degradation of the electrodes and an increased safety risk. In addition, a high ECR or bulk resistance lowers the power capability and efficiency of the total battery system.

Researchers measured the ECR of different battery cell connectors in the past years. Taheri *et al.* (2011) showed that the ECR between the electrode tabs of a pouch cell and the current-collector bar can be reduced to $10\,\text{m}\Omega$ with a sufficient contact force. As contact materials they use copper and brass. Brand *et al.* (2015) investigated the joints of the welded battery cell connections. Therefore, they applied resistance spot, ultrasonic and laser beam welding. The obtained ECR, the heat input and the ultimate tensile force of the joints were compared against each other for the application of brass test samples. The properties of brass enabled the process for all three welding techniques. Figure 3.16 summarized the results of the measured ECR.

Table 3.2 Benefits and drawbacks of clamped and welded joints based on the analyses of Lee et al. (2010).

Method	Advantage	Disadvantage
Resistance welding	• Low-cost • Efficient and fully automated • No filler metals or gases • Existing technology for weld quality control	• Difficult to produce large nuggets • Difficult for highly conductive materials • Electrode sticking/wear • Possible expulsions
Laser welding	• Non contact process • Less thermal input, less distortion • Very high precision of welding • High speed	• Hard to produce a large joint area • Needs good joint fit-up • Material reflectivity • Need of shielding gas • High initial cost
Ultrasonic welding	• Solid state process • Excellent for highly conductive materials • Good for thin sheets or wires • Good for multi-layered sheets • Gauge ratio insensitive • Dissimilar materials • No filler metals or gases	• Restricted to lap joints • Limited in joint thickness (<3 mm) • Challenging on high strength, hardness materials • Sensitive to surface conditions • Possible audible noise • Large weld indentation • Sonotrode sticking
Soldering	• Joining metals with different melting temperatures • High mechanical strength	• Heat input into battery cell • Additional filler metal and flux are necessary
Clamped contacts	• No heat input • Separable joint • Simple joining of different materials	• Additional parts and mass • Labor intensive • Mechanical stress limits the range of applicable cells

Brand et al. (2017) also investigated the ECR and tensile strength of the soldered brass samples. In comparison to the welding processes, the contact elements dit not melt during the soldering process. Instead, the two contact materials were joined together by a filler metal (solder). A heat source, which depended on the soldering technique, melted the solder between the contact interface, while the contact materials remained in the solid phase due to a higher melting point. A chemical reaction between the contact materials and the solder formed an intermetallic compound which showed a strong metallic bonding [Tu (2007), S. 16]. This enabled the joining of metals with different melting temperatures. The investigated joint between the brass samples showed a high mechanical strength and a low ECR [Brand et al. (2017)].

3.5.3 Clamped cell connectors

In order to establish clamped cell connectors for cylindrical 18650 battery cells, a similar ECR compared to welded and soldered cell connectors is necessary. Therefore, Bolsinger et al. (2017) investigated the ECR between different contact elements and the positive terminal of a Panasonic NCR18650B battery cell. A mechanical stress test assured that the positive terminal could be loaded with a force of up to 400 N without undergoing a plastic deformation. This enabled further investigations in which five different materials with three different surface roughnesses were clamped on the battery terminal with a variable force between 25 and 400 N. The influence of aged surfaces on the ECR was additionally investigated.

Figure 3.14 shows the schematic of the contact interface. Due to the asperities of the contact members, overlaying oxide and contaminant films, only a small portion of the apparent contact area A_a leads the electrical current through the interface, which is also called the conducting contact area A_r. The area, which is loaded but isolated with a dielectric film, is described by the load-bearing area A_{lb}. The conducting contact area consists of many asperities which constrict the current flow and therefore increases the resistance at the interface (constriction resistance). Thus, the total ECR consists of the constriction resistance and the film resistance. If a force is applied on the contact members the conducting area deforms, increases and creates further contact spots. Thus, the ECR can be reduced with an increasing contact force. Bolsinger et al. introduced the following equation which describes the contact resistance R_C. It is calculated with the specific resistance of the two contact members ρ_1 and ρ_2, the hardness of the softer contact member H, the applied contact force F, the specific film resistance ρ_c and the thickness of the film d_c to

$$R_C = \frac{(\rho_1 + \rho_2)\sqrt{\pi H}}{4k} F^{-0.5} + \frac{\rho_c d_c H}{\pi k} F^{-1} = r_s F^{-0.5} + r_f F^{-1}. \quad (3.1)$$

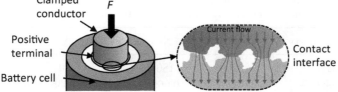

Fig. 3.14 Schematic of the contact interface between the battery terminal and the clamped conductor. The current only flows through some metallic contacts, which produce an additional constriction resistance.

The factor k considers the proportion from the conducting contact area A_r to the load bearing area A_{lb}. The factor r_s represents the part of constriction resistance whereas the factor r_f represents the part of the film resistance.

The positive terminal of the Panasonic NCR18650B battery cell offers a clamping area with a diameter of 7 mm. In order to validate the influences of the material and the surface, Bolsinger et al. used contact elements with different Brinell hardness, specific resistance and surface roughness. Table 3.3 lists the utilized contact elements. The considered materials are stainless steel (CrNiMoTi), nickel (Ni), brass (CuZn), aluminum (Al) and copper (Cu). For the experiment, the cylindrical contact elements with a diameter of 7 mm are placed on the positive terminal which consists of nickel plated steel. Then the contact force is increased from 25 to 400 N, while the ECR is measured.

Figure 3.15 shows the influence of the material constants on the measured ECR. The surface was therefore polished by a glasspaper with a grit size of 180. The ECR for all contact elements decreases with increasing contact force. The CrNiMoTi contact element shows the highest ECR over the total range of the applied force. Aluminum has the second-largest ECR. The oxide layer on the aluminum is responsible for the high gradients of the ECR over the applied force in comparison to the other contact elements. Copper shows the lowest ECR over the total range of the applied force. At the applied force of $F = 400$ N, all contact elements — except CrNiMoTi — show an ECR of $R_C < 100\,\mu\Omega$. The lowest ECR is reached for the copper element at $F = 400$ N with around $R_C = 50\,\mu\Omega$. Furthermore, measurements with a variated roughness, aging and diameter of the surfaces of the contact elements validate their influence on the ECR. The results show that the material constants like the Brinell hardness HB and specific resistance ρ are

Table 3.3 Contact elements that Bolsinger et al. used for ECR measurements on a positive battery cell terminal. The table lists the material properties, Brinell hardness HB, specific resistance ρ and the averaged surface roughness R_a.

Property	CrNiMoTi	Ni	CuZn	Al	Cu
HB	200	180	105	70	40
$\rho[\Omega \frac{mm^2}{m}]$	0.75	0.069	0.066	0.035	0.017
$R_{a,80}[\mu m]$	2.420	3.102	3.378	4.043	5.125
$R_{a,180}[\mu m]$	0.875	1.563	1.265	1.986	1.395
$R_{a,800}[\mu m]$	0.408	0.928	1.157	1.415	1.272

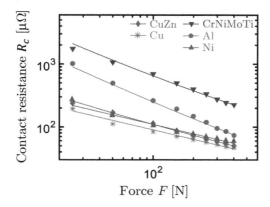

Fig. 3.15 Contact resistance between different conductors and the positive terminal of a 18650 battery cell. The measured resistances show the influence of the contact material and the applied force F. Each contact resistance decreases with increasing force F. The minimum contact resistance is measured at $F = 400\,\text{N}$ with copper as the contact material with $R_c = 50\,\mu\Omega$. Values according to Bolsinger et al. (2017).

the most important factors of influence on the ECR. Due to high applied forces, which breakup the oxide/film layer, aged surfaces have no crucial effect at forces of $F = 400\,\text{N}$ on the ECR [Bolsinger et al. (2017)].

3.5.4 *Conclusion*

The most important advantage of a clamped cell connector is the simple disassembling process that leaves no damage. The investigations of Bolsinger and colleagues show that a positive terminal of a cylindrical battery cell (Panasonic NCR18650B) can be loaded up to a force of $F = 400\,\text{N}$ without undergoing plastic deformation and is therefore applicable for clamped cell connectors. They also identify the hardness $H\text{B}$ and the specific resistance ρ of the contact elements, as well as the applied force as the most crucial parameters for designing a clamped contact. With regard to the ECR, Ni, Cu, CuZn and Al are all suitable contact materials. Furthermore, the standard electrode potential of the contact elements has to be considered in order to avoid galvanic corrosion. The ECR of the clamped cell connectors are in the same range or even lower (see Cu for $F = 400\,\text{N}$) than the welded and soldered cell connectors, which are investigated by Brand et al. (2017). Further investigations have to validate the long term stability related to the ECR of clamped cell connectors. Figure 3.16 compared the ECR of the different joining techniques.

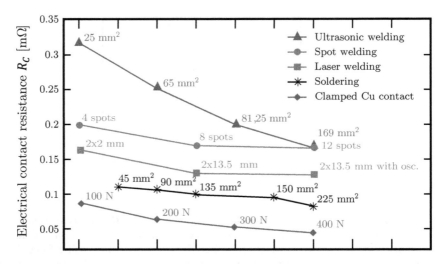

Fig. 3.16 Electrical contact resistances of different joining technologies for lithium battery cells. The electrical contact resistance of an ultrasonic welded joint depends on the area of the sonotrode. The electrical contact resistance of a spot welded joint depends on the number of welded spots. The ECR of a laser welded joint is plotted over the number and length of the welded lines. The soldering joint depends on the contact area. The electrical contact resistance of the clamped joint is depicted for a cylindrical copper conductor with a diameter of 7 mm and varied force between 100 N and 400 N. The values are according to Brand *et al.* (2017) and Bolsinger *et al.* (2017).

3.6 Battery Thermal Management

The temperature strongly affects several parameters of the lithium-ion cells. An increased temperature lowers the internal resistance of the battery cell and therefore increases the power capability and capacity, though it also accelerates aging processes. If the temperature exceeds a value of about 80–90°C the solid electrolyte interface (SEI) layer decomposes. This increases the safety risk. A decreased temperature increases the internal resistance and results in an increased capacity fade. In addition, lithium plating can be caused by charging at low temperatures, which will also increase the safety risk and aging processes. Also, the arrangement of single cells in a module will cause an uneven temperature distribution, which will lead to a localized deterioration. Achieving a maximum performance and cycle life of the cells require the integration of a battery thermal management (BTM). The crucial requirements to an ideal BTM are:

- Keeping the cell temperatures in the optimum operating range — cooling and heating.

- Reducing the temperature gradients between the cells in a module and pack.
- Reducing the temperature gradients within a cell.
- A compact and lightweight design.
- Simple integration in the battery pack or module.
- Reliable, low-cost and easy for servicing and maintenance.
- A provision for gas channels if a battery cell is venting.
- Low parasitic power consumption.

The BTM can be realized by different approaches and separated into two categories — the passive BTM (e.g., cooling by the ambient air) and the active BTM (e.g., cooling and heating by auxiliary heater). Passive BTMs do not use special heating or cooling units. They only use the ambient or the cabin air. To keep the cells at the optimum operating temperature, the ambient air needs to be in the range of 10–30°C. If the ambient air temperature is outside of these conditions, active components are necessary. Section 3.6.1 describes the different approaches for BTMs in detail. Section 3.6.2 shows the approach of the BTM for the modular battery with integral single cells.

3.6.1 State of the art

The BTMs use different systems and can be categorized into mediums that are utilized for cooling/heating. Air and liquid can be used for cooling or heating the battery cells. Furthermore, an insulation as well as a thermal storage (e.g., phase change materials), or even a combination of both can be used for temperature control. As with any system, the approach depends on several parameters like the available space, weight, complexity and costs. The following sections describe the three main approaches.

3.6.1.1 Air cooling for BTM

A simple way of cooling is the production of a parallel or serial air flow across the battery cells. The air that is used for cooling/heating is either direct ambient air, the conditioned air from the vehicle cabin or the air conditioned from a separate heater. Figure 3.17 shows the general principles of a BTM using air. The Toyota Prius (2001 model) and the Honda Insight (2000 model) for example, use the cabin air for cooling their battery packs [Kelly and Rajagopalan (2001), S.60].

Due to the air stream through the cells, the temperature of the air is higher at the outlet than at the inlet. This temperature gradient is also shown

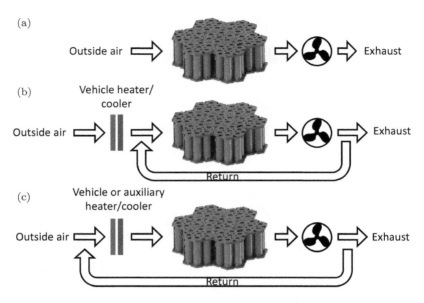

Fig. 3.17 General principles of thermal management using air. (a) shows the passive cooling with outside air. Therefore a fan generates the airflow that streams through the cells or through the fins of a cooling plate. (b) depicts the passive heating and cooling principle using the cabin air. (c) shows the active cooling and heating principle. Therefore, the outside air is heated or cooled by an auxiliary heat/cooler. The principles according to Pesaran (2001).

for the battery cells. Mahamud and colleagues produced a reciprocating air flow in a battery module consisting of cylindrical cells and created a much more uniform temperature distribution. Wang and Ma (2017) applied the method to a module consisting of 18 prismatic cells. They showed that the maximum temperature difference could be reduced to under 5°C. Wang et al. (2014) investigated different cell arrangements as well as an optimized fan location. They discovered that the best cooling performance was achieved by locating the fan on the top of the module. Chen et al. (2017) used this same fan location to improve the inlet and outlet structure by a respective divergence or convergence plenum. The optimization of the plenum widths reduced the temperature distribution by a further 45%.

If the battery is below a temperature of −10°C, it has to be heated up rapidly for the start of the vehicle. Furthermore, power optimized lithium-ion cells generate a huge amount of heat and therefore has to be cooled down rapidly. However, this is difficult to achieve by an air cooling system, due to the low thermal conductivity. Additionally, the lower thermal conductivity makes air cooling more ineffective compared to cooling by liquid. Designing an effective forced-air cooling system requires the optimization of the length and cross sectional area of the air stream path, the conditioned air temperature, the speed of airflow, the fan location and the arrangement of the battery cells [Wang et al. (2014)].

3.6.1.2 Liquid cooling for BTM

For large battery modules with compact cell arrangements, air cooling by natural and forced convection cannot sufficiently dissipate the heat from the cells. Compared to air, liquid has a higher thermal conductivity and higher heat capacity, therefore resulting in better cooling performance. Utilized mediums can be refrigerants or coolants like water, oil or acetone. Figure 3.18(a)–(c) depicts the general principles of thermal management of a vehicle battery. While cooling with refrigerant works without heating elements and additional loops for the chiller, it still cannot heat up the battery in sub-zero conditions. However, there exist different approaches for cooling the battery cells with liquid [Pesaran (2001)]:

- Discrete tubing around the battery/module
- Jacket around the cells/module
- Submerging cells/modules in a dielectric fluid for direct contact
- Use of cooling plates

Similar to the air flow, the structure and geometry of the liquid channels for cooling plates are optimized to produce an uniform temperature distribution. Jarrett and Kim (2011) investigated the influence of the channel routing, width and length on the coolant pressure drop, temperature uniformity of the cooling plate and the average temperature. They discovered that the design for optimized temperature uniformity had a narrow inlet channel widening towards the outlet. Zhao et al. (2015) designed a new cooling method for cylindrical battery cells based on mini-channel liquid cooled cylinders. Therefore, the battery cells are placed in the cylinders which are connected to flow distribution plates at their end caps. Anthony et al. (2017a) investigated cylindrical battery cells with integrated heat pipes. Compared to surface-based cooling methods, the heat pipe cooling resulted in a more uniform temperature distribution within the cell. However, the heat pipe insertion is a big manufacturing challenge.

If the fluid is not dielectric, sealing the cooling plate, tubing or jacket is essential. Although, the heat transfer coefficient of the indirect liquids (e.g., water, glycol) is higher in case of higher thermal conductivity and lower viscosity, the effectiveness of indirect cooling decreases due to the additional walls of the jacket, container or fins [Pesaran (2001)].

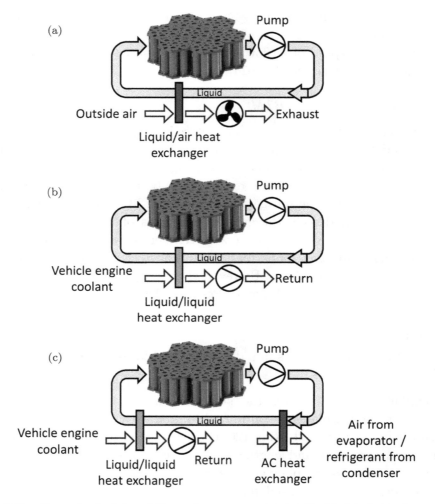

Fig. 3.18 General principles of thermal management using liquid. (a) shows the passive cooling principle using the outside air to cool the liquid/air heat exchanger. The liquid is pumped either directly through the battery cells or indirectly using jackets or cooling plates. In (b) a liquid/liquid heat exchanger is used to cool or heat the liquid with the vehicle engine coolant. (c) shows an active cooling and heating principle using an additional AC heat exchanger besides the liquid/liquid heat exchanger.

3.6.1.3 Phase change materials for BTM

Besides the air and liquid cooled systems, which require additional mass through blowers, tubing or pumps, PCMs can also be applied to cool batteries. Hallaj and Selman (2000) investigated the PCM for an automotive battery pack for the first time. The PCM stores the heat as latent heat that is generated from the battery cells during discharge. The stored heat is rejected to the battery during the relaxation or charging process when the temperature drops. In sub-zero regions, this is an advantage for the application in EVs. Hallaj and Selman's (2000) simulation show that with PCMs the temperature difference is more uniform than it is without. Furthermore, they defined the ideal PCM with the following requirements:

- Melting point between 30 and 60°C
- High latent heat per unit mass
- Narrow melting temperature range
- Inert with respect to other battery components
- Cheap, light, thermally cyclable, safe and nontoxic

Due to low thermal conductivity, the PCMs show a poor heat dissipation. Khateeb et al. (2004) showed that an additional aluminum foam increases the thermal conductivity by one magnitude, therefore significantly improving the heat dissipation. Further approaches to increase the thermal conductivity include "the embedding of a metal matrix into the PCM, the impregnating of porous materials, the adding of high thermal conductivity substances and the developing of latent heat thermal energy storage systems with unfinned and finned structures" [Wang et al. (2016)].

3.6.1.4 Heat pipe

Conventional cooling methods (see Secs. 3.6.1.1 and 3.6.1.2) use air or liquid to cool the surface of the battery cell. However, the heat in the core of the cell could not be effectively cooled due to the low thermal conductivity. Therefore, integrating a heat pipe into the cylindrical cell may be an effective cooling method. The heat pipe is a passive device that effectively transfers heat between two solid interfaces by using thermal conductivity and phase

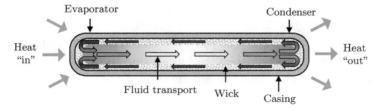

Fig. 3.19 Principal function of a heat pipe. The working fluid evaporates by absorbing the hot side. The vapor travels along to the cold side and condenses to liquid again. Thus, the latent heat is released. Capillary action, gravity or centrifugal force takes the fluid back to the evaporator.

transitions. Figure 3.19 shows the working principle of a heat pipe. The working fluid evaporates by absorbing the heat from the hot side of the heat pipe. The vapor than travels along to the cold side of the heat pipe and condenses to liquid again by releasing the latent heat. Capillary action, gravity or centrifugal force takes the fluid back to the hot side of the heat pipe and therefore closes the loop. The thermal conductivity of a heat pipe can be about 120 times higher as compared to copper.

Shah et al. (2016) integrated a heat pipe into the axis of the cylindrical cell and achieved an effective cooling with a 18 to 20°C temperature reduction in the cell core. They also observed that the utilization of a thin metal rod instead of a heat pipe shows similar results. Anthony et al. (2017a) also inserted a heat pipe into a cylindrical 26650 battery cell and compared the results with traditional surface-based cooling approaches. They showed that with the heat pipe insertion the core temperature becomes as low as, or even slightly lower than the outer surface, therefore improving the thermal uniformity in the cell. However, the heat pipe integration shows several technical challenges, such as the assembly process, interference with the cell connectors at the terminals and heat dissipation at the condenser side [Shah et al. (2016); Anthony et al. (2017a)]. Furthermore, the passive heat pipe is not able to produce heat for the use of the battery cell in sub-zero conditions.

3.6.1.5 *Thermoelectric cooler (TEC)*

Thermoelectric coolers (TEC) are solid state devices, thus they are reliable energy converters that show no noise or vibrations. They are small in size and are lightweight. This enables it to be used in a large number of applications. Through the Peltier effect, the devices convert the electrical energy into a thermal gradient [Riffat and Ma (2003)]. Depending on the DC voltage across two dissimilar materials, the temperature gradient could be regulated or even reversed. Therefore, TECs could be used to heat and cool the battery. Figure 3.20 shows the principal structure of a TEC. The ceramic material on both sides of the TEC enables direct contact, even with energized components like battery terminals due to the electrical insulation. The following list summarizes the advantages of the TEC [Alaoui (2013)]:

- No moving parts
- Small size and lightweight
- Maintenance free
- No noise

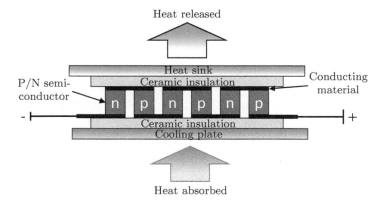

Fig. 3.20 Structure of a thermoelectric cooler. The DC current is applied to the conducting material and then flows through one or more pairs of n- and p-type semiconductor materials. Ceramic plates usually electrically insulate the conducting materials from the overlaying cooling plate or heat sink [Riffat and Ma (2003)].

- Heating and cooling with the same TEC
- Wide operating temperature range
- Highly precise temperature control
- Operation in any orientation, zero gravity
- Environmentally friendly
- Cooling to very low temperatures

Alaoui (2013) designed and evaluated TEC modules of a BTM for 60 Ah lithium-ion pouch cells. The BTM was then installed in an EV and tested with the US06 driving cycle. Furthermore, the cells were discharged by an constant current of 3 C, whereby the temperature was decreased from 57.69°C to 46.8°C by the TECs. The BTMs consumed 919 Wh, which corresponds to 9.8% of the total energy of the battery pack. However, the weight of the battery pack increased by 64.4% and its volume increased by 152.8%, due to the heat sinks with blowers that were necessary for absorbing the power generated at the TEC. Alaoui (2013) concluded that the key factor for an ideal operation of TECs for BTMs is the optimization of the heat sinks in terms of weight and thermal resistance.

In a further investigation Alaoui measured the performance and energy consumption of a BTM based on TECs for a 20-Ah lithium-iron-phosphate battery cell. The TECs were mounted on the hot spots of the cell which were at its center and at the cell terminals. A control circuit regulated each TEC independently to keep the cells in the operating temperature range between

−10°C and 40°C. Alaoui (2017) observed that under extreme conditions with an ambient temperature of 40°C and a discharge current of 3 C, the BTM consumed 41% of the total energy of the pack to keep the temperature of the cells in the operating range.

Alaoui and Salameh (2005) also measured the temperature and coefficient of performance (COP) for a BTM based on TECs integrated in vehicles. The COP describes the ratio of the cooling effect divided by the energy input. They observed a temperature of 52°C in the heating mode versus a temperature of 9.5°C in the cooling mode. The COP in the heating mode was 0.65 and 0.23 for the cooling mode. Thus, this system is much more effective for heating the battery.

Despite the many advantages of the application of TECs for BTMs, they show a low efficiency in the cooling mode. Due to the hot and cold sides of the TECs, direct integration between two prismatic battery cells is not possible. Otherwise, one cell will be cooled, with the other heated up. Therefore, the hot side of the TEC has to be connected to a heat sink.

3.6.2 BTM for integral single cell

3.6.2.1 Non-uniform temperature distribution inside battery cells

Besides the non-uniform temperature distribution in the battery pack or module, a non-uniform temperature distribution inside the battery cells exists as well. The temperature difference between the surface and the core of the cell depends on several parameters like the applied cooling method (heat dissipation), the thickness and geometry of the cell, the cell materials (thermal conductivity and heat capacity) and the current profile (heat generation). Zhang et al. (2014) investigated the radial temperature distribution of a cylindrical 18650 battery cell under various currents and cooling conditions. They measured a temperature difference between the core and surface of the cell — ∼3°C with natural convection cooling and ∼5.5°C with forced convection cooling. The temperature gradient increases with increasing currents and stronger cooling conditions. Temperature measurements of the cylindrical 26650 battery cells even show gradients of 10–15°C between the core and surface temperature [Forgez et al. (2010); Anthony et al. (2017a,b)]. Figure 3.21(a) and (b) shows the typical temperature distribution of a pouch and a cylindrical battery cell. Figure 3.21(c) depicts the core and the surface temperature of a cylindrical 26650 battery cell for a 20 A charge and discharge pulses [Forgez et al.

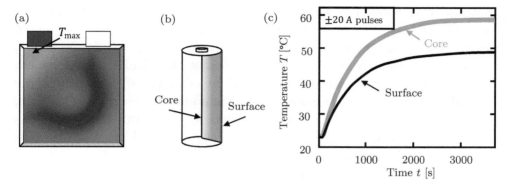

Fig. 3.21 Non-uniform temperature distribution over (a) a pouch and (b) within a cylindrical battery cell. Due to the higher electrical resistance and lower thermal conductivity of the cathode current collector (Al), compared to the anode current collector (Cu), the highest temperature occurs near the cathode terminal of a pouch cell [Bazinski and Wang (2014)]. The increased temperature in the center of a cylindrical battery cell is caused by the low thermal conductivity in the radial direction. (c) shows a temperature difference of about 10°C between the core and surface of a cylindrical 26650 battery cell for 20 A current pulses. Data according to Forgez et al. (2010).

(2010)]. The inhomogeneous temperature inside the cell results in uneven aging. Waldmann and Wohlfahrt-Mehrens (2015) used the integrated Arrhenius law and estimated the changes of concentration inside the cylindrical 18650 battery cell to be about 3 times higher in the core compared to the areas close to the surface. Klett et al. (2014) aged cylindrical 26650 battery cells in three different ways — (i) Using a simulated drive cycle, (ii) using a constant current cycle, and (iii) storing for three years. The investigated jelly roll of the cycled cells showed highly uneven aging between their center and outer areas. They assumed that the uneven degradation was a result from the temperature and pressure gradients. Thermal gradients also exist inside lithium-ion pouch cells [Hunt et al. (2016); Bazinski and Wang (2014)]. Therefore, an ideal BTM not only equalizes the gradients in the module, but also inside the battery cells.

3.6.2.2 Terminal cooling

Producing an equalized temperature distribution within the cells during the charge and discharge process, different cooling methods are investigated. Though various cooling configurations exist for lithium-ion pouch cells and prismatic hardcase cells, this is not the case for cylindrical battery cells [Chen et al. (2016); Hunt et al. (2016); Bazinski and Wang (2014)].

Bazinski and Wang (2014) investigated the thermal effect of cooling the cathode grid tabs of lithium pouch cells. The cathode tab was chosen because of its significantly higher temperature as compared to the anode. However, results showed that the tab cooling only had minimal benefit in reducing the temperature gradients. Hunt *et al.* (2016) investigated the degradation of tab and surface-cooled lithium-ion pouch cells. Therefore, the cells were tested using different currents for a total of 1000 cycles. As the current increased, the surface-cooled cells degraded more strongly, as compared to the tab-cooled cells. For discharge rates of 6 C the surface-cooled and tab-cooled cells showed a capacity fade of 9.2% and 1.2%, respectively. They deduced that this effect was caused by the orthotropic thermal conductivity of the cell. By surface cooling the parallel layers of the battery cell are at different temperatures, which leads to non-uniform impedance and non-uniform current distribution between the layer. However, tab cooling causes a uniform temperature distribution between each parallel layer.

The common cooling configuration of cylindrical battery cells removes the heat from the surface (can) by a forced air or liquid flow, phase change materials or jacket cooling. Due to the low thermal conductivity in a radial direction, a significant temperature gradient between the core and surface of the cell exists. Anthony *et al.* (2017a) removed the heat at the center of a cylindrical cell by integrating a heat pipe. This heat pipe reduced the temperature gradient between the core and the surface from about 13°C to about 1°C. Furthermore, the overall temperature of the battery cell was reduced effectively. However, the insertion of a heat pipe into a battery cell presents several significant manufacturing challenges.

For our modular battery we also used the mechanical and electrical interface between the battery cell and the positive/negative plate for cooling. Thus, the positive and negative plates can be cooled by several approaches, like an overlaying cooling plate or cooling fins (see Fig. 3.22(a)). The cooled positive and negative plates transfer the temperature directly to the respective positive or negative terminal of the battery cell where the highest temperature at the outside (surface) of the cell can be measured. This not only removes the heat which is produced in the cell connector by ohmic losses, but also dissipates the heat from the core of the cell through the multiple tabs. Figure 3.22(b) shows the jelly roll with their current collector tabs and the resulting heat transfer. This approach effectively reduces the temperature gradients as well as the overall temperature.

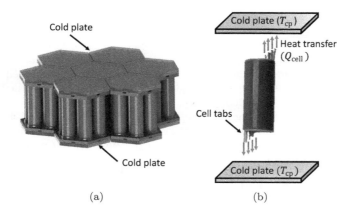

Fig. 3.22 Principle of terminal cooling for cylindrical battery cells. (a) shows the submodule which is cooled by two cooling plates placed on the positive and the negative plates. This cools the battery terminals as well as the current carrying plates. (b) depicts the jelly roll of a cylindrical battery cell with cell tabs. The heat transfer from the center of the cell to the cold plates occurs through the cell tabs. This principle uses the higher thermal conductivity in the axial direction compared to the radial direction.

Acknowledgment

The work presented in this book chapter is gathered within the research project LIBELLE, funded by the VECTOR STIFTUNG. The authors thankfully acknowledge the financial support.

Bibliography

Alaoui, C. (2013). Solid-state thermal management for lithium-ion EV batteries, *IEEE Trans. Veh. Technol.* **62**, 1, pp. 98–107.

Alaoui, C. and Salameh, Z. M. (2005). A novel thermal management for electric and hybrid vehicles, *IEEE Trans. Veh. Technol.* **54**, 2, pp. 468–476, http://www.freepatentsonline.com/8214097.html.

Alaoui, C. (2017). Modular energy efficient and solid-state battery thermal management system, *2017 Int. Conf. Wirel. Technol. Embed. Intell. Syst. WITS 2017*, pp. 1–5, doi:10.1109/WITS.2017.7934608.

Anthony, D., Wong, D., Wetz, D., and Jain, A. (2017a). Improved thermal performance of a Li-ion cell through heat pipe insertion, *J. Electrochem. Soc.* **164**, 6, pp. A961–A967, doi:10.1149/2.0191706jes, http://jes.ecsdl.org/lookup/doi/10.1149/2.0191706jes.

Anthony, D., Wong, D., Wetz, D., and Jain, A. (2017b). Non-invasive measurement of internal temperature of a cylindrical Li-ion cell during high-rate discharge, *Int. J. Heat Mass Transf.* **111**, pp. 223–231, doi:10.1016/j.ijheatmasstransfer.2017.03.095.

Bazinski, S. J. and Wang, X. (2014). Thermal effect of cooling the cathode grid tabs of a lithium-ion pouch cell, *J. Electrochem. Soc.* **161**, 14, pp. A2168–A2174, doi: 10.1149/2.0731414jes, http://jes.ecsdl.org/cgi/doi/10.1149/2.0731414jes.

Bolsinger, C., Zorn, M., and Birke, K. P. (2017). Electrical contact resistance measurements of clamped battery cell connectors for cylindrical 18650 battery cells, *J. Energy Storage* **12**, pp. 29–36, doi:10.1016/j.est.2017.04.001, http://linkinghub.elsevier.com/retrieve/pii/S2352152X16301529.

Brand, M. J., Kolp, E. I., Berg, P., Bach, T., Schmidt, P., and Jossen, A. (2017). Electrical resistances of soldered battery cell connections, *J. Energy Storage* **12**, pp. 45–54, doi:10.1016/j.est.2017.03.019, http://linkinghub.elsevier.com/retrieve/pii/S2352152X16303310.

Brand, M. J., Schmidt, P. A., Zaeh, M. F., and Jossen, A. (2015). Welding techniques for battery cells and resulting electrical contact resistances, *J. Energy Storage* **1**, 1, pp. 7–14, doi:10.1016/j.est.2015.04.001, http://dx.doi.org/10.1016/j.est.2015.04.001.

Chen, D., Jiang, J., Kim, G. H., Yang, C., and Pesaran, A. (2016). Comparison of different cooling methods for lithium-ion battery cells, *Appl. Therm. Eng.* **94**, pp. 846–854, doi:10.1016/j.applthermaleng.2015.10.015, http://dx.doi.org/10.1016/j.applthermaleng.2015.10.015.

Chen, K., Wang, S., Song, M., and Chen, L. (2017). Structure optimization of parallel air-cooled battery thermal management system, *Int. J. Heat Mass Transf.* **111**, pp. 943–952, doi:10.1016/j.ijheatmasstransfer.2017.04.026.

Daowd, M., Omar, N., van den Bossche, P., and van Mierlo, J. (2011). A review of passive and active battery balancing based on MATLAB/Simulink, *Int. Rev. Electr. Eng.* **6**, 7, pp. 2974–2989, doi:10.1109/VPPC.2011.6043010.

Forgez, C., Vinh Do, D., Friedrich, G., Morcrette, M., and Delacourt, C. (2010). Thermal modeling of a cylindrical LiFePO4/graphite lithium-ion battery, *J. Power Sources* **195**, 9, pp. 2961–2968, doi:10.1016/j.jpowsour.2009.10.105, http://linkinghub.elsevier.com/retrieve/pii/S037877530901982X.

Hallaj, S. A. and Selman, J. R. (2000). A novel thermal management system for electric vehicle batteries using phase-change material, *J. Electrochem. Soc.* **147**, 9, p. 3231, doi:10.1149/1.1393888, http://jes.ecsdl.org/cgi/doi/10.1149/1.1393888.

Hunt, I. A., Zhao, Y., Patel, Y., and Offer, J. (2016). Surface cooling causes accelerated degradation compared to tab cooling for lithium-ion pouch cells, *J. Electrochem. Soc.* **163**, 9, pp. A1846–A1852, doi:10.1149/2.0361609jes, http://jes.ecsdl.org/lookup/doi/10.1149/2.0361609jes.

Jarrett, A. and Kim, I. Y. (2011). Design optimization of electric vehicle battery cooling plates for thermal performance, *J. Power Sources* **196**, 23, pp. 10359–10368, doi: 10.1016/j.jpowsour.2011.06.090, http://dx.doi.org/10.1016/j.jpowsour.2011.06.090.

Kelly, K. J. and Rajagopalan, A. (2001). Benchmarking of OEM hybrid electric vehicles at NREL, August, pp. 1–102, doi:NREL/TP-540-31086.

Ketterer, B., Karl, U., Moest, D., and Ulrich, S. (2009). Lithium-ion batteries: State of the art and application potential in hybrid-, plug-in hybrid- and electric vehicles, *Forschungszentrum Karlsruhe Wissenschaftliche Berichte FZKA 7503*, p. 22.

Khateeb, S. A., Farid, M. M., Selman, J. R., and Al-Hallaj, S. (2004). Design and simulation of a lithium-ion battery with a phase change material thermal management system for an electric scooter, *J. Power Sources* **128**, 2, pp. 292–307, doi: 10.1016/j.jpowsour.2003.09.070.

Klett, M., Eriksson, R., Groot, J., Svens, P., Ciosek Högström, K., Lindström, R. W., Berg, H., Gustafson, T., Lindbergh, G., and Edström, K. (2014). Non-uniform aging of cycled commercial LiFePO4//graphite cylindrical cells revealed by post-mortem analysis, *J. Power Sources* **257**, pp. 126–137, doi:10.1016/j.jpowsour.2014.01.105, http://dx.doi.org/10.1016/j.jpowsour.2014.01.105.

Lee, S. S., Kim, T. H., Hu, S. J., Cai, W. W., and Abell, J. A. (2010). Joining technologies for automotive lithium-ion battery manufacturing — A review, *ASME 2010 Int. Manuf. Sci. Eng. Conf. MSEC 2010*, 12 to 15 October 2010, **1**, 4, pp. 541–549, doi:10.1115/MSEC2010-34168.

Pesaran, A. (2001). Battery thermal management in EVs and HEVs: Issues and solutions, *Adv. Automot. Batter. Conf.*, p. 10.

Riffat, S. B. and Ma, X. (2003). Thermoelectrics: A review of present and potential applications, *Appl. Therm. Eng.* **23**, 8, pp. 913–935, doi:10.1016/S1359-4311(03) 00012-7, arXiv:1512.00567.

Ruiz, V., Pfrang, A., Kriston, A., Omar, N., Van den Bossche, P., and Boon-Brett, L. (2017). A review of international abuse testing standards and regulations for lithium-ion batteries in electric and hybrid electric vehicles, *Renew. Sustain. Energy Rev.*, June 2016, pp. 1–26, doi:10.1016/j.rser.2017.05.195, http://dx.doi.org/10.1016/j.rser.2017.05.195.

Shah, K., McKee, C., Chalise, D., and Jain, A. (2016). Experimental and numerical investigation of core cooling of Li-ion cells using heat pipes, *Energy* **113**, pp. 852–860, doi:10.1016/j.energy.2016.07.076.

Taheri, P., Hsieh, S., and Bahrami, M. (2011). Investigating electrical contact resistance losses in lithium-ion battery assemblies for hybrid and electric vehicles, *J. Power Sources* **196**, 15, pp. 6525–6533, doi:10.1016/j.jpowsour.2011.03.056, http://dx.doi.org/10.1016/j.jpowsour.2011.03.056.

Tu, K.-N. (2007). *Solder Joint Technology*, Vol. 92, 1st Edition (Springer-Verlag New York), ISBN 978-0-387-38890-8, doi:10.1007/978-0-387-38892-2, http://www.springerlink.com/index/10.1007/978-0-387-38892-2%5Cnhttp://link.springer.com/10.1007/978-0-387-38892-2.

Waldmann, T. and Wohlfahrt-Mehrens, M. (2015). In-operando measurement of temperature gradients in cylindrical lithium-ion cells during high-current discharge, *ECS Electrochem. Lett.* **4**, 1, pp. A1–A3, doi:10.1149/2.0031501eel, http://eel.ecsdl.org/content/4/1/A1.abstract.

Wang, H. and Ma, L. (2017). Thermal management of a large prismatic battery pack based on reciprocating flow and active control, *Int. J. Heat Mass Transf.* **115**, pp. 296–303, doi:10.1016/j.ijheatmasstransfer.2017.07.060.

Wang, Q., Jiang, B., Li, B., and Yan, Y. (2016). A critical review of thermal management models and solutions of lithium-ion batteries for the development of pure electric vehicles, *Renew. Sustain. Energy Rev.* **64**, pp. 106–128, doi:10.1016/j.rser.2016.05.033.

Wang, T., Tseng, K. J., Zhao, J., and Wei, Z. (2014). Thermal investigation of lithium-ion battery module with different cell arrangement structures and forced air-cooling strategies, *Appl. Energy* **134**, pp. 229–238, doi:10.1016/j.apenergy.2014.08.013, http://dx.doi.org/10.1016/j.apenergy.2014.08.013.

Zhang, G., Cao, L., Ge, S., Wang, C.-Y., Shaffer, C. E., and Rahn, C. D. (2014). In situ measurement of radial temperature distributions in cylindrical Li-ion cells, *J. Electrochem. Soc.* **161**, 10, pp. A1499–A1507, doi:10.1149/2.0051410jes, http://jes.ecsdl.org/cgi/doi/10.1149/2.0051410jes.

Zhao, J., Rao, Z., and Li, Y. (2015). Thermal performance of mini-channel liquid cooled cylinder based battery thermal management for cylindrical lithium-ion power battery, *Energy Convers. Manag.* **103**, pp. 157–165, doi:10.1016/j.enconman.2015.06.056, http://dx.doi.org/10.1016/j.enconman.2015.06.056.

Zhu, J., Zhang, X., Sahraei, E., and Wierzbicki, T. (2016). Deformation and failure mechanisms of 18650 battery cells under axial compression, *J. Power Sources* **336**, pp. 332–340, doi:10.1016/j.jpowsour.2016.10.064, http://dx.doi.org/10.1016/j.jpowsour.2016.10.064.

Chapter 4

Parallel Connection of Lithium-ion Cells — Purpose, Tasks and Challenges

Alexander Fill

4.1 Introduction

Lithium-ion batteries are the favorite energy supplier in many applications, such as home energy storage, mobile applications, smart grids and electric vehicles. Because of different requirements the battery design has to change significantly for each application. Beneath the power and energy amount, the battery has to fulfill safety, lifetime, reliability and cost requirements. One option to manage these requirements is the connection of the cells in the battery.

In order to estimate the needed current/load and voltage levels, the required power and energy can be taken into account. The power is given by the product of voltage and current $P = U \cdot I$ and the energy amount is the integral of power over the time, which can be expressed with the product of nominal voltage and capacity $E = U \cdot Q$. Whereby the voltage level is the sum of all serial-connected cell voltages and the capacity is the sum of all parallel-connected cell capacities. The maximal current is to a certain extent related to the sum of current limits of all parallel-connected cells, which will be discussed later.

The battery cost increases stepwise with increasing voltage and current levels, due to assembly limits and higher demand on isolation and current rails. Also, laws and regulations change at different voltage levels, e.g., test and safety instructions. Therefore, both parameters have to be changed in order to reach the power and energy requirements.

In order to achieve the required energy, the number of parallel cells needed is the ratio of the needed capacity to the capacity of one cell

$n_{parallel} = C_{battery}/C_{cell}$. So the number of parallel cells can, depending on the cell size, theoretically vary from one up to infinity. Safety,[1] efficiency and lifetime issues of the battery will be affected by the cell size and the number of parallel-connected cells. These will be discussed further in this chapter.

4.2 Main Issues and Challenges

The lifetime of battery cells is a key issue for the commercialization of lithium-ion batteries in many applications. In order to ensure the longest possible lifetime, the cells have to be operated in specific ranges, e.g., temperature, voltage and current ranges. These ranges are mostly correlated to each other and/or to other cell characteristics, e.g., the State of Charge (SoC)[2] or State of Health (SoH).

The battery management system (BMS) has to ensure that the battery operates optimally. In order to do so it is necessary to measure or accurately predict all cell parameters and characteristics. In order to compare parallel- and serial-connected cells, the focus is set on the current determination.

Given by the first Kirchhoff's law the currents of each serial-connected cell in a string are equal. This is, however, not valid for parallel-connected cells, as illustrated in Fig. 4.1. Because of cost and space issues cell currents cannot be measured for each cell. The unknown current distribution complicates the prediction of the cell's SoC and SoH and leads also to gradients in temperature, discharge level and exposed stress within the logical cell. This results in varying cell aging and affects the battery's lifetime and reliability.

Therefore the main task for operating parallel-connected cells is an accurate prediction of the current distribution. The response of temperature, SoC and SoH to the current distribution intensifies this issue and complicates the

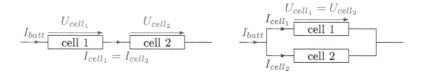

Fig. 4.1 Comparison of the current distribution for serial- and parallel-connected cells.

---

[1] Safety issues will be discussed in Chapter 8.
[2] The SoC is abbreviated by κ in equations in this chapter.

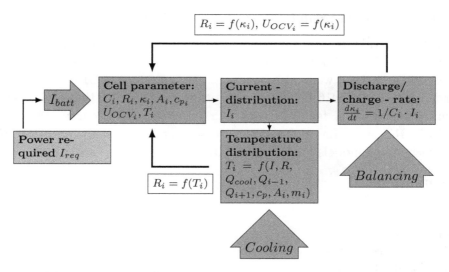

Fig. 4.2 Schematic model of parameters and the BMS's influence on the current distribution.

current prediction. Figure 4.2 shows the responsibilities of the most crucial parameters and the influences the BMS can take. The parameter correlations are generating a nonlinear differential equation system, which makes the issue non trivial. In order to predict the current distribution, correlations to other measurable parameters have to be found. The dissipation energy, for example, increases almost by the power of two with increasing current difference of parallel-connected cells. This is shown by Eq. (4.1), which gives the dissipation energy of two parallel cells, with the assumption of equal cell resistances $R_1 = R_2 = R_{Cell}$

$$q_{dissipation} = R_{Cell} \cdot \delta I^2 + 2(I_1^2 - I_1 \delta I), \tag{4.1}$$

with the current of cell 1 and the current difference of cell 1 to cell 2, $\delta I = I_1 - I_2$.

4.3 Influences on the Current Distribution

In this section, the influences of the most relevant parameters on the current distribution and their possible consequences will be discussed. Therefore, a simplified model will be introduced that offers an analytical solution for the cell currents, followed by more realistic assumptions to help evaluate, e.g., load changes, influences of the open circuit voltage (OCV) curve and thermal effects.

4.3.1 Simplified model — Analytical solution

In order to evaluate the influences on the current distribution of parallel-connected cells, a simplified model with an analytical solution was made. These assumptions were also made in [Brand *et al.* (2016) and Pastor-Fernandez *et al.* (2016)]. There were other publications that used similar assumptions with a few additions, e.g., realistic OCV curves [An *et al.* (2016)], RC-pairs for polarization [Gong *et al.* (2015)] and thermal-electrochemical models [Yang *et al.* (2016)].

Assumptions:[3]

(1) Parallel-connection of two cells
(2) No polarization
(3) Cell consists of U_{OCV} and inner resistance
(4) Constant inner resistance $R = const.$
(5) Constant capacity $C = const.$
(6) Constant cell temperature $T = const.$
(7) Equal cell temperature $T_1 = T_2$
(8) Linear OCV curve $U_{OCV} = x_0 \kappa + x_1$
(9) Constant battery load $I_{batt} = const.$
(10) No degradation

In order to describe the cell characteristics R_1 and C_1 are introduced, which represent the inner resistance and the capacity of cell one. The resistance and the capacity of cell two are given as ratios to cell one: $x_R = R_2/R_1$ and $x_C = C_2/C_1$. The values for the cell characteristics are chosen such that:

- Resistance ratio $x_R = R_2/R_1$ with $x_R \epsilon [1, \infty)$
- Capacity ratio $x_C = C_2/C_1$ with $x_C \epsilon (-\infty, \infty)$

The current of cell one is derived by transposing the first Kirchhoff's law.

$$I_1 = \frac{\Delta_{2-1} U_{OCV}}{R_1(1 + x_R)} + \frac{x_R}{1 + x_R} I_{batt}, \qquad (4.2)$$

with the OCV difference of cell 2 to cell 1 $\Delta_{2-1} U_{OCV}$ and the battery current I_{batt}. The SoC of the cells and the time derivative of the SoC difference of

---

[3] These assumptions, though not clearly elaborated upon, are still valid for the whole chapter.

cell 2 to cell 1 are displayed in Eqs. (4.3) and (4.4).

$$\kappa_i = 1/C_i \int I \, dt, \tag{4.3}$$

$$\frac{\delta \Delta_{2-1} \kappa}{\delta t} = \frac{1}{C_1 x_C}[I_{batt} - I_1(1 + x_C)], \tag{4.4}$$

with the SoC difference of cell 2 to cell 1 $\Delta_{2-1}\kappa$. The time derivative of the OCV difference of cell 2 to cell 1 can be described by Eq. (4.5).

$$\frac{\delta \Delta_{2-1} U_{OCV}}{\delta t} = \frac{1}{C_1 x_C}[I_{batt} - I_1(1 + x_C)]x_0, \tag{4.5}$$

with the slope of the OCV curve x_0. Fitting Eq. (4.5) in Eq. (4.2) and subsequently integrating with the boundary condition $I_1 = x_R/(1 + x_R)I_{batt}$ at $t = 0^4$ results in Eq. (4.6), which gives the time dependency of the current of cell one.

$$I_1(t) = \frac{1}{x_C + 1} I_{batt} \left[\left(x_R \frac{x_C + 1}{x_R + 1} - 1 \right) e^{-\frac{t}{\tau}} + 1 \right], \tag{4.6}$$

with the time constant τ, given in Eq. (4.9). So the current distribution of two parallel-connected cells can be described with a few characteristic values demonstrated in Fig. 4.3 and Eqs. (4.7) to (4.11).

As can be seen from Fig. 4.3 the uneven cell currents, at the beginning of the discharge, cause a difference in the discharge rate. The discharge rate generates an unequal SoC and leads to an unequal OCV level. The time derivative of the OCV difference changes the current distribution as described in Eq. (4.2). The cell currents run exponential to an even discharge rate, while the SoC and OCV differences run exponential to their limits given by Eqs. (4.10) and (4.11). Brand et al. (2016) was the only publication that theoretically describes the influences of x_R and x_C on the current distribution of parallel-connected cells. They obtained the same results with confirming experiments, but this was not discussed in detail.

$$\frac{\Delta_{1-2} I_{t \to \infty}}{I_{batt}} = \frac{1 - x_C}{1 + x_C}, \tag{4.7}$$

$$\frac{\Delta_{1-2} I_{t=0}}{I_{batt}} = \frac{x_R - 1}{x_R + 1}, \tag{4.8}$$

---

[4]It was assumed that the SoC of the cells are equal at the beginning of the discharge.

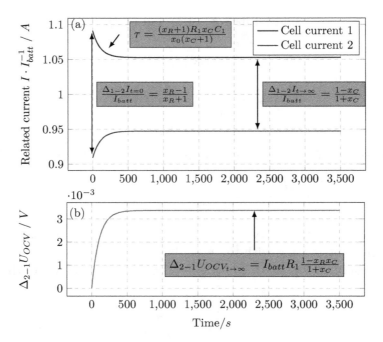

Fig. 4.3 Analytical solution of the current distribution (a) and the OCV difference (b), with $x_R = 1.2$, $x_C = 0.9$, $R_1 = 0.001\ \Omega$, $C_1 = 40$ Ah and $I_{batt} = 1C$.

$$\tau = \frac{(x_R + 1)R_1 x_C C_1}{x_0(x_C + 1)}, \qquad (4.9)$$

$$\Delta_{2-1}U_{OCV_{t\to\infty}} = I_{batt}R_1 \frac{1 - x_R x_C}{1 + x_C}, \qquad (4.10)$$

$$\Delta_{2-1}\kappa_{t\to\infty} = I_{batt}R_1 \frac{1 - x_R x_C}{(1 + x_C)x_0}. \qquad (4.11)$$

Impacts of x_R and x_C With the discussed equations the influences of the resistance and capacity ratios, x_R and x_C, respectively, on the current distribution can be evaluated. Three cases can be differed:

(1) $x_R \cdot x_C = 1$
(2) $x_R \cdot x_C < 1$
(3) $x_R \cdot x_C > 1$

Case (1) represents a balanced system. The current distribution caused by the resistance ratio corresponds to the final distribution with even discharge rates. In this case the currents will be constant and no differences in SoC and OCV will arise between the cells.

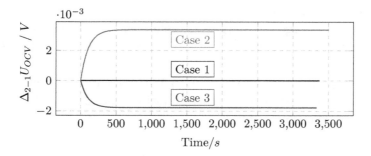

Fig. 4.4 OCV difference $\Delta_{2-1}U_{OCV}$ for the three cases: Case (1) with $x_R = 1.2$ and $x_C = 0.833$; case (2) with $x_R = 1.2$ and $x_C = 0.8$ and case (3) with $x_R = 1.2$ and $x_C = 0.9$.

Cases (2) and (3) differ in a higher distribution — case (2) at the beginning and case (3) at the end of the discharge. This affects the value of the OCV and SoC differences, see Fig. 4.4. These values are crucial for current distribution at the point of battery current steps. A more detailed view is given in the next paragraph.

It is clear from Eq. (4.2) that a negative OCV difference $\Delta_{2-1}U_{OCV}$ at the point of discharge leads to an additional current difference of the parallel cells, when it is instantly switched to charging.[5] In order to evaluate this, case (3) has to be subdivided:

3.1 $x_R > 1, x_C < 1$
3.2 $x_R > 1, x_C > 1$

Case (3.1) represents typical different aged cells. In this case the possible maximal current difference is smaller as the current difference is caused by x_R only.[6]

A higher current difference at the point of instantly switching from discharge to charge, can appear in case (3.2). This case can arise by inducing a higher temperature in the "worse" cell. Therefore, thermal management is a main issue for parallel-connected cells, which will be discussed later.

Impacts of I_{batt} In order to show the influence of current changes two effects will be discussed. First a stepwise current increase/decrease

---

[5]The effect is identical for the switch from charging to discharging with a positive $\Delta_{2-1}U_{OCV}$.
[6]The fact that $x_R > x_R \cdot x_C$ causes a lower OCV difference, that leads to lower current differences.

to the required current and second, an instant switch of the battery current from discharge to charge. Therefore, a parameter $x_I = I_{batt}/I_{1C}$[7] is introduced, which gives the battery current related to the 1C charge current.

Equation (4.2) can be extended to battery current profiles by transforming the boundary conditions. Equations (4.12) to (4.15) give the corresponding equation system.[8]

$$I_1(t) = I_{stable}\left[\left(\frac{I_{start}}{I_{stable}} - 1\right)e^{-\frac{t}{\tau}} + 1\right], \qquad (4.12)$$

$$I_{start_i} = \frac{\Delta_{2-1}U_{OCV_i}}{R_1(x_R+1)} + \frac{x_R}{x_R+1}I_{1C}x_{I_i}, \qquad (4.13)$$

$$\Delta_{2-1}U_{OCV_i} = I_{end_{i-1}}R_1(x_R+1) - R_1 x_R I_{1C} x_{I_{i-1}}, \qquad (4.14)$$

$$I_{stable_i} = \frac{I_{1C}x_{I_i}}{1+x_C}. \qquad (4.15)$$

In Fig. 4.5 the related current and the OCV difference for different step numbers, with the same summarized step time Fig. 4.5(a) and the impacts of the step time Fig. 4.5(b) are shown. In order to compare the results the corresponding currents and voltages are time shifted.

The steps lead to a difference of the OCV between the cells, which react on the final current distribution as a "time shift" x_τ. The "time shift" increases with decreasing difference between the actual and final OCV difference. The shift can be increased with increasing step time and decreasing gaps between the final step and the required current. The effects can be used to limit the maximal current difference within a logical cell.

The current distribution for battery current profiles was theoretical discussed in [Brand et al. (2016)] and experimentally shown in [Gogoana et al. (2014); Bruen and James (2016);[9] Fouchard and Taylor (1987)]. The measured currents showed the same behavior as illustrated here.

The cell currents can be calculated in the same way for the scenario of instant switching from discharge to charge. The focus was set to the maximal current difference's "worse" that can appear. This is reached, if the battery current is switched at the point of maximal OCV difference and can be

---

[7]I_{1C} represents the current needed for a full battery charge after one hour.
[8]The indices i and $i-1$ represent the actual and prior current state.
[9]In [Bruen and James (2016)], the current distribution for four cells at current changes was researched by experiment and simulation.

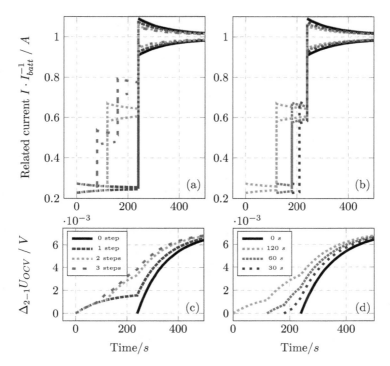

Fig. 4.5 Influences of the current change factor x_I; impacts of the step number with summed step time 240 s on the current distribution (a) and OCV difference (c): 3 steps with $x_I = 0.25$, $x_I = 0.5$ and $x_I = 0.75$, 2 steps with $x_I = 0.25$ and $x_I = 0.5$, 1 step with $x_I = 0.25$; impacts of step time current on the distribution (b) and OCV difference (d) with 2 steps (right): 240 s, 120 s and 60 s step times.

calculated as follows:

$$\frac{\Delta_{1-2} I_{max}}{I_{batt}} = 2 \frac{1 - x_R x_C}{(1 + x_C)(1 + x_R)} \frac{x_{I_{i-1}}}{x_{I_i}} + \frac{x_R - 1}{x_R + 1}. \qquad (4.16)$$

This equation is valid for all battery current steps. It can be seen that the issue is only relevant for $x_R \cdot x_C > 1$. Two possible effects should be discussed. First, a cell current higher than the battery current, and second, the maximal possible current differences within a logical cell, which corresponds to $x_R \to \infty$.

In order to estimate the needed resistance ratio x_R, an equal capacity $x_C = 1$ and equal absolute currents $x_{I_{i-1}} = -x_{I_i}$ are assumed. This effect only appears at high ratios for x_R and x_C. Aging causes contrary ratios and that is why thermal effects are assumed. The temperature mainly affects the resistance ratio, therefore $x_R/x_C \gg 1$ and x_C is set to one. For these assumptions the theoretical resistance ratio has to be higher than three.

The possibility of cell resistance ratios higher than three have to be evaluated by measurements with commercially used cells. The limit of the current difference can be predicted using the same assumption and $x_R \to \infty$ to $\Delta_{1-2} I_{limit}/I_{batt} = 3/2$.

4.3.2 Effects of cell resistance and capacity variations

Variations on cell parameters can have many origins and reasons, but in principle can be divided into two groups: (1) variations caused by production, e.g., inner resistance, capacity and self discharge rate, and (2) gradients caused by operation, e.g., temperature, SoC and aging gradients. Here, the effects from the first group, which are impossible to prevent, will be discussed. The variation caused by production is experimentally shown in many publications, e.g., [Gong et al. (2015); Schuster et al. (2015); Brand et al. (2016); Gogoana et al. (2014); Kim and Cho (2013)]. Most of these publications demonstrated the influences on current distribution after connecting the cells, with highest parameter differences, in a parallel configuration.

Here, a statistical analysis for a more general view was made. The focus was set on the effects of the number of parallel-connected cells. Therefore, the used model will be discussed first.

Electrical model In order to simulate n parallel-connected cells, an equivalent circuit model (ECM) with the above assumptions was built and transformed, as shown in Fig. 4.6. The transformation reduces the system to $n-1$ $2p$ connections. The OCV and the inner resistance have to be transformed as described in Fig. 4.6. A recursion can be used to constitute any parallel-connection in one code.

Simulation In order to simulate the variations of cell resistance and capacity a normal distribution was assumed, see Fig. 4.7. Although the distribution of each parameter varies, especially because of the different cell capacities, it was assumed that the cell distribution for all simulations was the same. This was done because the relation between distribution and production and cell size is not known. In addition, the distribution will vary with production level, cell type and cell producer.

In order to demonstrate the influence of the number of parallel-connected cells on the current distribution a Monte Carlo simulation[10] for 2, 3, 5, 10

---

[10]For the Monte Carlo simulation, 1e6 single simulations for each parallel-connection were done.

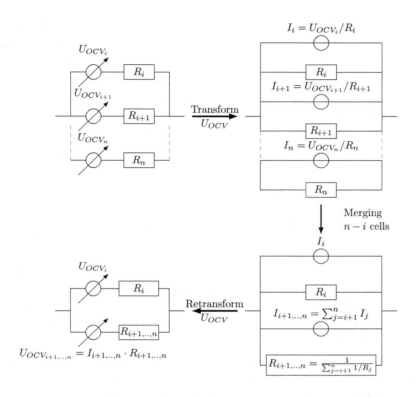

Fig. 4.6 Creating an equivalent circuit model with the above-mentioned descriptions for n parallel-connected cells and transforming it to an equivalent $2p$ connection.

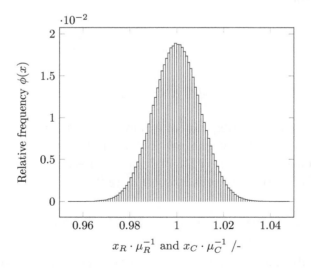

Fig. 4.7 Assumed cell resistance x_R and capacity x_C distribution related to their expected values μ_R and μ_C and a related standard deviation of $\sigma \cdot \mu^{-1} = 0.01$.

and 100 parallel-connected cells was done. A constant discharge of $1C$ and constant related standard deviations for resistance and capacity distributions $\sigma_R \cdot \mu_R^{-1} = \sigma_C \cdot \mu_C^{-1} = 1e-2$ were conducted. The expected cell resistances and capacities were adapted to the number of parallel cells n, $\mu_R = 1e-3 \cdot n$ Ω and $\mu_C = 120 \cdot n^{-1}$, in order to keep logical cell resistances and capacities equal for the different configurations.

In order to compare the results a current distribution factor λ_I and SoC distribution factor λ_{SoC}, see Eqs. (4.17) and (4.18), were introduced.

$$\lambda_I = \left(\frac{I_j \cdot n}{I_{batt}}\right)_{max}, \quad j \epsilon \{1,...,n\}, \qquad (4.17)$$

$$\lambda_{SoC} = \left(\frac{\sum_{i=1}^{n} (\kappa_i - \kappa_j) x_{C_i}}{\sum_{i=1}^{n} x_{C_i}}\right)_{max}, \quad j \epsilon \{1,...,n\}. \qquad (4.18)$$

Results The results of the Monte Carlo simulations are displayed in Fig. 4.8. The figure shows the influences of the number of parallel cells n on the distribution of the current distribution factor at load start $\lambda_{I,start}$ (Fig. 4.8(a)), final limit $\lambda_{I,final}$ (Fig. 4.8(b)), the SoC distribution factor λ_{SoC} (Fig. 4.8(c)) and the time constant τ (Fig. 4.8(d)).

The distribution of the current distribution factor λ_I shifts to higher values with increasing number of parallel cells n, as demonstrated in Fig. 4.8(a) and (b). This can be explained by a mathematical description of the current distribution factor λ_I with cell resistance and capacity distributions x_R and x_C.

$$\lambda_{I,start} = \left(\frac{n}{x_{R_j} \sum_{i \epsilon \{1,...,n\} \setminus \{j\}} \frac{1}{x_{R_i}} + 1}\right)_{max}, \qquad (4.19)$$

$$\lambda_{I,final} = \left(\frac{n \cdot x_{C_j}}{\sum_{i=1}^{n} x_{C_i}}\right)_{max}. \qquad (4.20)$$

The shift can therefore be explained by a higher distribution of the maximum cell value of x_R and x_C, which can be calculated with the corresponding probability density function (PDF) of the cell resistance x_R, capacities x_C and the number of parallel cells n, see Eq. (4.21).

$$\phi(x_{G,C})_{max} = PDF(x_{G,C}) \cdot \left(\int_{-\infty}^{\infty} PDF(x_{G,C}) d(x_{G,C})\right)^{n-1}. \qquad (4.21)$$

The SoC distribution factor λ_{SoC} can also be expressed with cell parameters x_R and x_C, see Eq. (4.22), which can be reduced to Eq. (4.23), if n

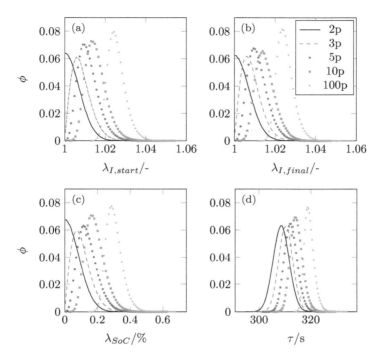

Fig. 4.8 Results of the Monte Carlo simulation for 2, 3, 5, 10 and 100 parallel-connected cells, with Gaussian distributed cell resistances x_R, capacities x_C and 1e6 simulations for each connection. Represented are the distributions of the current distribution factor at load start (a), final limit (b), SoC distribution factor (c) and time constant (d). The parameters of the related standard deviations, expected resistance and capacity, and battery current were set to $\sigma_R \cdot \mu_R^{-1} = 1e-2$, $\sigma_C \cdot \mu_C^{-1} = 1e-2$, $\mu_R = 0.001 \cdot n\ \Omega$, $\mu_C = 120 \cdot n^{-1}$ Ah and $I_{batt} = 1C$.

approaches infinity

$$\lambda_{SoC} = \frac{I_{batt}\mu_R n}{m_{OCV}\left(\sum_{i=1}^{n} x_{C_i}\right)^2} \sum_{j=1}^{n} \left((x_{R_j}x_{C_j})_{min} - x_{R_i}x_{C_i}\right)x_{C_i}, \quad j\epsilon\{1,\ldots,n\}, \tag{4.22}$$

$$\lambda_{SoC,n\to\infty} = \frac{I_{batt}\mu_R}{m_{OCV}}\left((x_{R_j}x_{C_j})_{min} - 1\right)x_{C_i}, \quad j\epsilon\{1,\ldots,n\}. \tag{4.23}$$

The minimum distribution of the product of cell resistance and capacity $x_R x_C$ shifts to lower values with an increasing number of parallel cells. This can also be calculated by Eq. (4.21), as resistance and capacity distributions were assumed to be independent.

Impacts on the time constant can be explained by extending Eq. (4.9) to n parallel cells. The time constant $\tau(n)$ reduces to Eq. (4.24), if n

approaches infinity

$$\tau_{i,n\to\infty} = \frac{x_{R_i} x_{C_i} \mu_R \mu_C}{m_{OCV}}. \tag{4.24}$$

The time constant correlates to the product of cell resistance and capacity. The cell parameter product $x_R x_C$ increases with a higher number of parallel cells n, which can be determined by Eq. (4.21), as already discussed.

Simplification — Extension of the analytical solution The current distribution can be analytically solved with the additional assumption of equal discharge rates of the merged cells in Fig. 4.6. With the above transformed model, the current distribution can be reduced to $n-1$ $2p$ connections. For the first $2p$ connection the current distribution is solved in Eq. (4.28).

$$x_{R_{2,..,n}} = \frac{1}{\sum_{i=2}^{n} \frac{1}{x_{R_i}}}, \tag{4.25}$$

$$I_1 = \frac{\Delta_{2-1} U_{OCV_{2,..,n-1}}}{R_1(1+x_{R_{2,..,n}})} + \frac{x_{R_{2,..,n}}}{1+x_{R_{2,..,n}}} I_{batt}, \tag{4.26}$$

$$\frac{\delta \Delta_{2-1} U_{OCV}}{\delta t} = \frac{1}{C_1(\sum_{i=2}^{n} x_{C_i})} \left[I_1 \left(1 + \left(\sum_{i=2}^{n} x_{C_i}\right)\right) - I_{batt} \right] x_0, \tag{4.27}$$

$$I_1(t) = \frac{I_{batt}}{1+\sum_{i=2}^{n} x_{C_i}} \left[\left(\frac{(1+\sum_{i=2}^{n} x_{C_i}) x_{R_{2,..,n}}}{1+x_{R_{2,..,n}}} - 1 \right) e^{-\frac{t}{\tau}} + 1 \right], \tag{4.28}$$

$$\tau = \frac{C_1(\sum_{i=2}^{n} x_{C_i})(1+x_{R_{2,..,n}}) R_1}{x_0(1+\sum_{i=2}^{n} x_{C_i})}. \tag{4.29}$$

4.3.3 *Influence of the open circuit voltage bending*

Experimentally-measured current distributions for a full discharge of parallel-connected cells showed a relation to the SoC's respective OCV level, see [Pastor-Fernandez et al. (2016); Yang et al. (2016)]. The currents typically reach a stable current distribution after a short time and cross each other at low SoCs with high current differences at the end of the discharge. It is assumed that the OCV bending causes this effect.

In order to investigate further, Assumption (4.3.1) in Sec. 4.3.1 was extended to become a quadratic relation and two new cases were discovered — a positive and negative bending of the OCV curve as given by Eqs. (4.30) and (4.31). They were chosen for two aspects: (1) the OCV curve differs for each active material,[11] but mostly consists of a part with a positive and a negative bend and (2) the assumption should show a quantitative influence of the OCV bend.

$$U_{OCV} = x_0 \kappa^2 + y_{OCV_1}, \quad (4.30)$$

$$U_{OCV} = -x_0(1-\kappa)^2 + y_{OCV_2}, \quad (4.31)$$

with the OCV at unload y_{OCV_1} and full-loaded state y_{OCV_2}. Figure 4.9 shows the related current distribution, OCV and SoC difference for the two

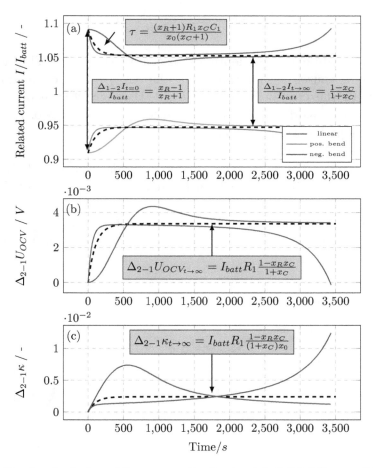

Fig. 4.9 Simulated solution of the current distribution (a), OCV difference (b) and SoC difference (c) for bended OCV curves, with $x_R = 1.2$, $x_C = 0.9$, $R_1 = 0.001\Omega$ and $C_1 = 40Ah$.

---

[11]The OCV curve can also differ for cells with the same active material.

cases at a 1C discharge. For comparison, the solution with linear OCV is added.

This figure shows relevant differences between the three cases. The influences of the OCV bending on the current distribution, OCV and SoC difference, demonstrated in Fig. 4.9 can be described qualitatively by the time derivative of the OCV difference of cell 2 to cell 1.

$$\frac{\delta \Delta_{2-1} U_{OCV}}{\delta t} = 2x_0 \left[\kappa_2 \frac{\delta \kappa_2}{\delta t} - \kappa_1 \frac{\delta \kappa_1}{\delta t} \right], \quad (4.32)$$

$$\frac{\delta \Delta_{2-1} U_{OCV}}{\delta t} = 2x_0 \left[\psi_2 \frac{\delta \kappa_2}{\delta t} - \psi_1 \frac{\delta \kappa_1}{\delta t} \right], \quad (4.33)$$

whereby ψ represents the depth of discharge (DOD).[12] The additional relation to κ causes a sign change in the $\delta \Delta U_{OCV}/\delta t$ curve during the discharge. In the case of a positive bend, the sign change occurs before ΔU_{OCV} reaches the discussed limit, because $\kappa_2 > \kappa_1$ for $t > 0$. I_1 rises with a positive slope limited by $I_1/I_{batt} = x_R/(x_R + 1)$.

The negative bend leads to a sign change after the discussed limit, because $\psi_1 > \psi_2$. The decreasing current I_1 leads to a second sign change and the current I_1 increases with a negative slope to the limit $I_1/I_{batt} = 1/(1 + x_C)$.

In order to evaluate these results, a simulation with an OCV curve consisting of three parts was made and compared with the measured current distribution, see Fig. 4.10.

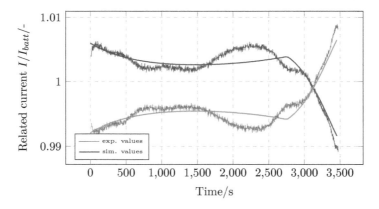

Fig. 4.10 Comparison between the experimental and the simulated current distribution.

---

[12]The depth of discharge is related to the state of discharge by: $\psi = 1 - \kappa$.

Caused by the more complex OCV curve the current distribution of the experiment and simulation differ, but the aim of Fig. 4.10 is to show the effects of the OCV bending.

In Fig. 4.10 the high current overshoot leads to the cell voltage limit, before the maximum of $\Delta_{2-1}U_{OCV}$ is reached. The overshoot is related to $\Delta\psi$ and the OCV slope. In other publications such as [Pastor-Fernandez et al. (2016); Yang et al. (2016)], it was experimentally found that cell currents converge after reaching a certain peak at the end of discharge, which confirms Fig. 4.9.

4.4 Thermal Effects

Thermal management will be one of the main tasks for parallel-connected cells. The uneven current distribution relates to the dissipative energy $p_{diss} = I^2 \cdot R_i$ and will mostly heat up the cell with the higher current. The relation of the inner resistance to the temperature can lead to a higher dispersion of the currents and effects described in Sec. 4.3.1. However, the thermal influences, e.g., heat flow of neighbor cells, different cooling and the current distribution self, on the current distribution and the development of temperature gradients in the battery has to be further researched.

Thermal influences on parallel-connected cells have not been frequently reported yet, since few publications cover all thermal effects, especially cooling, heat flux of neighbor cells and effects of current profiles. Their focus was mainly on accelerated aging caused by uneven temperatures.

The influence of uneven cell temperatures on the current distribution was researched in [Yang et al. (2016)]. Two cells[13] were connected in parallel, whereby the temperatures of the cells were controlled by a temperature chamber. The current distribution at a $1C$ discharge for different temperature gaps T_{gap}[14] and different ambient temperatures T_{amb} was researched.

The maximum current increased from $1.1 \cdot I_{batt}$ to $1.2 \cdot I_{batt}$ by increasing T_{gap} from $5\ K$ to $10\ K$. This is caused by the correlation of inner resistance and cell temperature. The current overshoot at the end of the discharge increased with T_{amb} and T_{gap}. The maximum current increased from $1.1 \cdot I_{batt}$ to $1.2 \cdot I_{batt}$ with $T_{amb} = 5°C$ to $25°C$ and from $1.2 \cdot I_{batt}$ to $1.4 \cdot I_{batt}$ with $T_{gap} = 5\ K$ to $10\ K$. This was induced by a higher dynamic caused by a lower resistance level. The influences of cooling, neighbor cells and load profiles were not researched.

---

[13] Yang et al. (2016) used cells with the 26650 format and $LiFePO_4$ as a cathode material.
[14] T_{gap} represents the temperature difference between the parallel-connected cells.

In [Pastor-Fernandez et al. (2016)], the temperature development of four parallel-connected cells[15] that had a different SoH level,[16] at a $1C$ discharge, was measured. Temperature differences of about 5 K between the cells arose at the end of the discharge. This temperature lay between the worst cell and a cell with median SoH level, whereby the first one had the higher temperature due to the high dissipation energy, caused by maximum cell currents at the end of the discharge.

In [Shi et al. (2016)], two parallel cells[17] were cycled by uneven temperatures[18] and compared with even tempered cells. The current distributions looked similar to Fig. 4.10, with higher current differences of the cells with uneven temperatures, especially at the end of the charge/discharge.

The publications show the influence of uneven temperatures on the current distribution and that temperature differences within the logical cell can arise by process. In order to evaluate the thermal influence, all thermal effects had to be considered as described above, especially since temperature development at the current profiles and how the temperature gradients in the battery can be controlled by cooling and load has to be researched.

4.5 Aging

As already mentioned the battery's lifetime will be a main issue for the commercialization of lithium-ion batteries. As the topic of aging will be addressed in another chapter, only the influences of the parallel-connection will be discussed. Therefore, a review of the present publications on this topic is given.

In [Pastor-Fernandez et al. (2016)], four pre-aged cells were connected in parallel and cycled 500 times. A cycling step consisted of a $1C$ discharge and a CCCV charge, with $0.5C$ and a current limit of $0.2C$. They discovered that SoH_p and SoH_E differences within a logical cell decreased upon further cycling, from 40% to 10% and 45% to 30%, respectively.

Shi et al. (2016) compared the capacity fade for two parallel-connected cells with and without temperature difference. The logical cells were cycled 1000 times by a $1C$ discharge and the charge was followed by 30 minutes

---

[15] Pastor-Fernandez et al. (2016) used 18650 cells with 3 Ah.
[16] The SoH was divided into SoH_E and SoH_p, which can be calculated with $SoH_E = (C_i - C_{eol})/(C_{bol} - C_{eol}) \cdot 100$ and $SoH_P = (R_i - R_{eol})/(R_{bol} - R_{eol}) \cdot 100$.
[17] Shi et al. (2016) used $LiFePO_4$ cells with a capacity of 60 Ah.
[18] The cell temperatures were kept by $25°C$ and $55°C$.

of rest. In order to evaluate the influence of the parallel-connection, two single cells were cycled with cell temperatures of 25°C and 55°C, respectively.

The cells that received the uneven temperatures aged two times faster than the ones that received even temperatures. A comparison with the single cells showed a similar aging result to the 25°C cell for the cells that received even temperatures. Cells with a temperature gap saw the lower temperature cell age faster almost in the way of the single cell by 55°C. After 1000 cycles, cells without a temperature gap achieved a 95% capacity retention rate, whereas for the cell pair with a temperature gap, the capacity retention rate was 90% and 85%, respectively.

Yang et al. (2016) presented the impact of uneven temperatures of two parallel-connected cells. In their experiment, two parallel-connected cells were cycled 1000 times using a $5C$ discharge, followed by CCCV charging with $3C$ to a current limit of $0.02C$. The capacity loss per cycle almost doubled from 0.3 to 0.5 mAh per cycle at 25°C ambient temperature by increasing the temperature gap from 0°C to 12°C. The effect was lower for an ambient temperature of 5°C, where the capacity loss increased from 0.1 to 0.15 mAh per cycle.

Gogoana et al. (2014) researched the influence of internal resistance mismatch of two parallel cells on the logical cell's lifetime. In their experiment, they connected in parallel, two cell pairs with low, middle and high resistance mismatch, before cycling them. Results showed that a mismatch of 20% caused a lifetime reduction of 40%, compared to cells with no resistance gap. Of critical importance was the use of only two measurements for each pair, and the influence of the resistance level. For the low resistance mismatch they used the cells with the lowest resistances. In order to validate this result, more cell pairs have to be tested using equal logical cell resistances.

Gong et al. (2015) characterized and tested cells that had already been used for two years in EVs. Cells with different capacity gaps were connected in parallel and the influence of the number of parallel cells on the current distribution was researched. Connections of $2p$ to $4p$ were considered, with the same maximum capacity gap of 10 Ah.

The results showed that a higher capacity gap led to increased current differences. A higher number of parallel-connected cells reduced the current differences within the logical cell. The maximum current ratio decreased from 4 to 3, then to 2.5 by increasing the number of cells in parallel from 2 to 3 then to 4. Knowing the unknown resistance force is critical, as it can be responsible for the differences in current.

Bibliography

An, F., Huang, J., Wang, C., Li, Z., Zhang, J., Wang, S., and Li, P. (2016). Cell sorting for parallel lithium-ion battery systems: Evaluation based on an electric circuit model, *J. Energy Storage* **6**, pp. 195–203.

Brand, M. J., Hofmann, M. H., Steinhardt, M., and Schuster, S. F. (2016). Current distribution within parallel-connected battery cells, *J. Power Sources* **36**, 334, pp. 202–212.

Bruen, T. and James, M. (2016). Modeling and experimental evaluation of parallel-connected lithium-ion cells for an electric vehicle battery system, *J. Power Sources* **310**, pp. 91–101.

Fouchard, D. and Taylor, J. (1987). The MOLICEL rechargeable lithium system: Multicell aspects, *J. Power Sources* **21**, pp. 195–205.

Gogoana, R., Pinson, M., Bazant, M., and Sarma, S. (2014). Internal resistance matching for parallel-connected lithium-ion cells and impacts on battery pack cycle life, *J. Power Sources* **252**, pp. 8–13.

Gong, X., Xiong, R., and Mi, C. (2015). Study of the characteristics of battery packs in electric vehicles with parallel-connected lithium-ion battery cells, *IEEE Trans. Ind. Appl.* **51**, pp. 1872–1879.

Kim, J. and Cho, B. (2013). Screening process-based modeling of the multicell battery string in series- and parallel-connections for high accuracy state-of-charge estimation, *J. Energy* **57**, pp. 581–599.

Pastor-Fernandez, C., Bruen, T., Widanage, W., Gama-Valdez, M., and Marco, J. (2016). A study of cell-to-cell interactions and degradation in parallel strings: Implications for the battery managment system, *J. Power Sources* **329**, pp. 574–585.

Schuster, S. F., Brand, M. J., Berg, P., Gleissenberger, M., and Jossen, A. (2015). Lithium-ion cell-to-cell variation during battery electric vehicle operation, *J. Power Sources* **297**, pp. 242–251.

Shi, W., Hu, X., Chao, J., Jiang, J., Zhang, Y., and Yip, T. (2016). Effects of imbalanced currents on large-format lifepo4/graphite batteries systems connected in parallel, *J. Power Sources* **313**, pp. 198–204.

Yang, N., Zhang, X., Shang, B., and Lia, G. (2016). Unbalanced discharging and aging due to temperature differences among the cells in a lithium-ion battery pack with parallel combination, *J. Energy Storage* **306**, pp. 733–741.

Chapter 5

Fundamental Aspects of Reconfigurable Batteries: Efficiency Enhancement and Lifetime Extension

Nejmeddine Bouchhima, Matthias Gossen and Kai Peter Birke

5.1 Introduction

Lithium-ion batteries are the most promising solution for energy storage systems in Battery Electrical Vehicles (BEVs) due to their low self-discharge rate and high-energy density. By interconnecting a large number of single cells, a reasonable kilometer range can be reached. On the other hand due to the manufacturing process tolerances, it is not possible to have identical cells regarding capacity and aging behavior in a multi-cell battery [Lehner et al. (2016); Paul et al. (2013)]. This cell imbalance gives rise to the degradation of the battery capacity and performance reduction over its lifetime, especially with regard to storage systems with serially connected cells [Lu et al. (2013)]. To address this issue, several balancing circuits are developed and optimized. Balancing topologies are generally classified in two types that are passive balancing and active balancing. While passive balancing dissipates the excess energy through shunt resistors in the form of heat, active balancing transfers charge from higher charged cells to lower charged cells [Daowd et al. (2013)]. In Gallardo-Lozano et al. (2014) a detailed overview of the active balancing method was provided. Since these systems are designed for the conventional battery architecture in which the cells are permanently connected to the load, as shown in Fig. 5.1(b), they are disabled to avoid any performance decrease by a high cell imbalance because of a single defective cell which causes the storage system to fail. To address this issue, a modular battery system

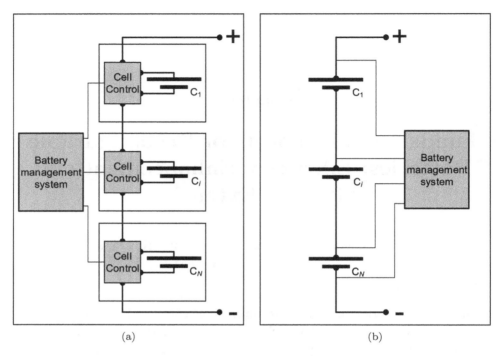

Fig. 5.1 (a) Reconfigurable battery architecture that comprises a battery managemnt system (BMS) and cell control units enabled to connect or disconnect the cell modeules, and (b) conventional battery architecture that consists of a number of cells permanently connected in series.

consisting of cell modules connected in series was proposed. Figure 5.1(a) illustrates this novel battery architecture.

While battery cells are decoupled, each of them is connected to a power control unit forming the cell module. The battery management system classifies these cells according to their actual State of Charge (SoC) and selects the best cells based on the implemented control strategy. At each instance of the battery operation the best module configuration is used to provide the power demand. These reconfigurable battery systems are widely discussed in the literature from a point of view hardware realization [Kim et al. (2011)]. For the supervisor control, rule-based strategies are almost implemented, such as the one presented in [Manenti et al. (2011)] where according to the control rules, the number of cells that could be deactivated is fixed regardless of the cell spread within the battery or the energy request.

In fact, the advantages of reconfigurable batteries depend on how the cell modules are controlled. Therefore, it is of great interest to develop the control strategy from the optimization theory's point of view maximizing profits, such as efficiency and lifetime enhancement gained by the novel battery architecture. To this end, the cell control is formulated as a non-linear dynamic optimization problem. Thereafter, to calculate the optimal discharge policy, an optimization algorithm using the dynamic programming techniques is being developed that will be able to achieve global optimum over the optimization horizon, e.g., a known driving cycle.

5.2 Modeling

The first step in the design of the optimal control strategy is to develop an appropriate mathematical model for this novel dynamic architecture derived from physical laws and logical relationship. The model should represent the system behavior depending on the phenomena of interest.

5.2.1 *Energy efficiency*

The energy efficiency of a storage system is expressed as

$$\eta_\mathrm{E} = \frac{E_\mathrm{el}}{E_\mathrm{bat}}, \tag{5.1}$$

where E_el is the electrical energy put into the load and E_bat is the energy stored in the battery. The flow of energy during one full discharge cycle can be depicted by a Sankey diagram, as shown in Fig. 5.2. This diagram summarizes the energy transfers taking place in a discharge process, thus self-discharge or capacity loss are not taken into account.

This Sankey diagram for a lithium-ion storage system shows that a part of the input energy is transferred as heat, while another part remains

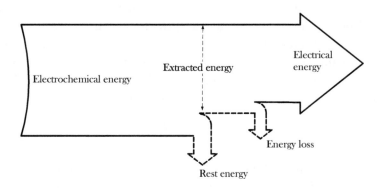

Fig. 5.2 Sankey diagram of the total energy flow during one discharge cycle.

unusable after the end of the discharge. These terms are described in detail on the previous page.

5.2.1.1 *Energy loss*

Energy loss is the amount of the chemical potential energy stored in the battery that is transformed into irreversible heat. This effect is attributed to overpotentials [Srinivasan and Wang (2003)], which are composed of ohmic overpotential ΔU_Ω and electrochemical reaction overpotentials ΔU_{Elc} [Dincer and Rosen (2007)]. Note that if the exchange current is high, as for electric vehicle batteries, then the electrochemical reaction overpotentials are relatively small in comparison to ohmic overpotential [Bard and Faulkner (2001)]. Thus, the ohmic heating will be taken into account by calculating the loss energy. This heating effect is due to internal resistance and can be described using Joule's law:

$$\Delta U_\Omega = R_{bat} \cdot i_{bat}. \tag{5.2}$$

5.2.1.2 *Rest energy versus equalization energy*

As discussed in the introduction, the cell mismatch in a conventional battery limits the battery capacity to the weakest cell. The remaining energy after the cut-off cannot be used and thus it is called rest energy. It is also possible to have a defective cell in the battery pack. In this worst case, the battery stops working and the energy content of the healthy cells cannot be used anymore.

In contrast to conventional storage systems, the cells within the reconfigurable batteries are dynamically disconnected based on their actual SoC, in the way that only the best n_{on} cells are selected to provide the power demand. By disabling n_{off} cells over the discharge process, the connected cells should deliver more electrical charges than if all battery cells N were connected. The amount of energy which is additionally delivered from the active cells during the battery discharge allows the equalization of the SoC among cells during the battery operation. Therefore, this energy is denoted as equalization energy E_{bal} and can be also expressed as

$$E_{bal} = n_{off} \cdot \frac{E_{bat}}{N}. \tag{5.3}$$

The purpose of the cell equalization is to discharge all battery cells in an appropriate way. Hence, the amount of rest energy depends on the battery equalization level, i.e., cells are partially or completely balanced, and reached

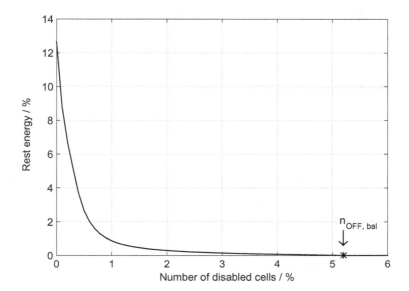

Fig. 5.3 The remaining energy in the battery at the end of a full discharge cycle as a function of the number of disabled cells (normalized to the number of battery cells).

at the end of the discharge. In fact, if a large number of battery cells is activated, the cells with low SoCs are fully discharged too quickly, resulting in incompletely balanced cells with unused remaining electrical charges. On the other hand, the larger the n_off, the better the cells are balanced, and thus there are fewer charges that remain in the battery after reaching the cut-off voltage. The evolution of the rest energy, denoted by E_res can be analytically described by a function $H\colon \mathbb{N} \to \mathbb{R}$ that defines the relationship between
$$n_\text{off} \to E_\text{res}$$
the remaining energy after reaching the cut-off voltage and the number of disabled cells during the discharge. Figure 5.3 shows the graph of the function H for an aged multi-cell storage system.

For $n_\text{off}=n_\text{off,Bal}$ no more electrical charges remain in the battery, which means that the cells are completely balanced and the whole storage energy is extracted. Thus, a further increase in n_off does not enhance the energy efficiency of the storage system.

5.3 Dynamic Optimization Problem

It is clear that disconnecting cells have two opposite effects on the energy efficiency of the battery. On one hand, decreasing the number of active cells reduces the amount of energy that remains in the battery after the end of discharge. On the other hand, it increases the battery current which results in a significant rise of the generated energy loss, as Joule heating depends on the square of the current. Therefore, in order to enhance battery efficiency, the profit obtained from disconnecting weak cells should be higher than the generated energy loss. To this end, the active number of cells at each instance should be optimized to let the battery attain the best possible energy efficiency over the optimization horizon. Due to its multistage nature, the optimization procedure can be well formulated as a discrete optimal control.

In general, an optimal control problem is described by an objective function, an equation introducing the system dynamics, and a control law and system constraints. In the following, the explicit definitions of these quantities are provided for the present optimal control:

1. Control law:
   The number of cells that should be activated at kth time step, $k = 0, \ldots, T - 1$, is the controllable input parameter in the present optimization problem. Therefore, this number represents the discrete control variable u_k. Due to the battery's current limitation and minimum required battery voltage, the control variable is a natural number constrained by the minimum number of active cells $u_{\min,k} = \max(n_{\max,\text{current}}, n_{\min,\text{voltage}})$ and the total number of battery cells N.

2. State equation:
   The state x_k shows the energy equalization level at kth time step. Therefore, the dynamic system is described by the following discrete-time state equation

   $$x_{k+1} = x_k + \left(1 - \frac{u_k}{N}\right) \cdot (\Delta t \cdot P_{\text{bat},k}), \qquad (5.4)$$

   where Δt is the time step duration and $P_{bat,k}$ is the battery power at kth time step.

   It is clear that at time $k = 0$ the equalization process still is not proceeding and thus $x_0 = 0$. Moreover, since the optimal control aims to maximize the battery efficiency by optimizing the cell equalization process, the end equalization level is also to be optimized. Accordingly, the terminal state x_T is free, but could not be higher than $x_T^{\max}$, which is the equalization level of fully balanced cells.

3. Performance criteria:
   The objective function describes the energy efficiency of the battery. As shown in Fig. 5.2, the amount of electrical energy put into the load

is the difference between the extracted energy and the energy loss. Therefore, the objective function $J(x_k, u_k)$ is expressed as

$$J(x_k, u_k) = F(x_T) - \sum_{k=0}^{T-1} g(P_{\text{bat},k}, u_k, \Delta t), \qquad (5.5)$$

where $g(P_{bat,k}, u_k, \Delta t)$ is the energy loss generated over Δt and $F(x_T)$ is the extracted energy with respect to the final equalization level x_T.

The discussed dynamic optimization problem consequently consists in calculating the control policy $\pi^* = \{u_0^*, \ldots, u_{T-1}^*\}$ that allows the objective function to reach a maximum value and at the same time satisfies the constraints listed below:

$$\begin{aligned}
\max_{u_k \in U_k} J(x_k, u_k) &= F(x_T) - \sum_{k=0}^{T-1} g(P_{\text{bat},k}, u_k, \Delta t) \\
\text{s.t.} \quad x_{k+1} &= x_k + f(x_k, u_k) \\
x_0 &= 0 \\
x_T &\leq x_T^{\max} \\
n_{\min,k} \leq u_k &\leq N \qquad \forall u_k \in U_k \\
k &= 0, 1, \ldots, T-1
\end{aligned} \qquad (5.6)$$

5.4 Optimal Control

This section presents the optimization algorithm to solve the optimal control problem addressed in this work. As the control variable u_k is an integer, the use of either conventional direct methods or indirect methods are not valid. Namely, these methods are based on the calculation of variations and evaluation of the Hamiltonian where the control variable is permitted to take any real value. A discussion of direct and indirect methods is given in von Stryk and Bulirsch (1992). Instead, this problem can be tackled with the use of dynamic programming method (DP), which is a powerful method for the numerical solution of dynamic optimization [Elbert et al. (2013)]. This method can deal with integer control variable and nonlinear dynamic and it is able to achieve global optimality.

5.4.1 *Vector-based dynamic programming*

To provide the global optimum of the studied objective function, it is mandatory to exhaustively enumerate the search space. In contrast to brute-force enumerations, the DP significantly reduces the computational cost of the optimal solution calculation by transforming the complex problem into a sequence of simpler problems. However, the conventional DP involves three loops over time, state variable and control variable. Replacing them through matrices will considerably improve the algorithm performance in terms of computation time. Therefore, the control variable is incremented from the minimum limit $n_{min,k}$ to N in single steps forming a grid of size $p_k \times T$. Thus, the control space at each step is presented by the vector $\boldsymbol{U_k} = [u_k^1, u_k^2, \ldots, u_k^i, \ldots, u_k^{p_k}]$ with $k = 0, \ldots, T-1$. Analogously, the state space is discretized to the discrete set $\{x_k^1, x_k^2, \ldots, x_k^j, \ldots, x_k^{d_k}\}$, forming a grid of size $d_k \times T$ over time and state. Note that x_k^j is an element of the state vector $\boldsymbol{X_k}$, and denotes the state variable in the discretized time-state space at the node with time index k and state index i. A discrete state x_k^j is calculated according the transition Eq. (5.4).

The introduction of a cost-to-go vector $\boldsymbol{G_k}$ is required in order to determine the optimal state. Each element G_k^i corresponds to the minimum energy loss generated when the state x_k^i is reached. Namely, as mentioned above, a state represents a battery equalization level which can be attained by distinct control policies, each generating a different amount of energy loss. In addition, a cost stage vector $\boldsymbol{g_k}$ is introduced, each of whose elements represent the energy loss generated by applying a control signal u_k^i over δt.

5.4.2 Complexity of the control strategy

A significant strength of the vector-based optimization algorithm is the reduction of the computing time. A performance evaluation and speed up factor calculation of vectorized code for a large state grid can be found in Dantas De Melo et al. (1990). However, a high memory allocation could be needed as the complexity of a classical DP is $O(T \cdot d^a \cdot p^b)$, where a is the number of d-dimensional states and b is the number of p-dimensional control variables. To address this issue, the state grid size is reduced in two ways, yielding the improvement of the proposed algorithm complexity. First, a power histogram is generated from the given driving cycle by associating the points of the driving cycle falling into the same power class. Note that the range of power values is broken into intervals, called class intervals or bins. A power class is presented by the average value $P_{\text{bat},k}$ of the power range. The number of associated points determines the number of occurrences of this bin, denoted by Δt_k. Thus, the number of the stages T, which corresponds now to the number of classes in the histogram, can be reduced without affecting the optimization result. Namely, the chronological order of power does not have a significant impact as the variation of cell parameters, such as internal resistance and open circuit voltage, with the SoC being described by a step function [Kim et al. (2013)]. Therefore, these parameters are assumed to be constant over one optimization horizon. The second measure is to keep at each stage, the state vector size as small as possible by deleting either duplicate elements or states that already violate the boundary conditions given in Eq. (5.6). Note that a significant improvement using the latter method can be achieved if the power histogram classes are classified in descending order with respect to their energy content $E_k = P_{\text{bat},k} \cdot \Delta t_k$. As a result, the number of states needed to calculate the optimal control policy is reduced to $O(c) = \sum_{k=0}^{T-1} \cdot d_k \cdot p_k$. To assess this complexity reduction a comparison between the proposed algorithm and the classical DP is conducted.

As can be seen from Fig. 5.4, the complexity of the proposed algorithm is least 10 times less than the one of the classical DP. Moreover, a significant reduction can be reached by decreasing the number of bins T in the power histogram. Note that the accuracy of the optimal policy obtained is highly dependent on the grid size T and on the discretization of the state space.

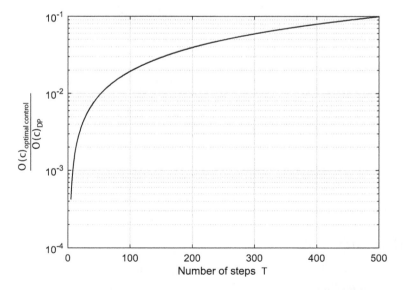

Fig. 5.4 Dependency of the algorithm complexity on the number of steps T. All computations are performed for a level of discretization by 10^4. The complexity of the basic DP is calculated with respect to the length of the optimization horizon.

Therefore, choosing a good level of discretization is a trade-off between accuracy and calculation time.

5.4.3 Optimal control policy

After calculating the vectors $\boldsymbol{G_k}$, $\boldsymbol{X_k}$ and $\boldsymbol{g_k}$, the optimization algorithm begins searching for an optimal solution. First, the optimal end state x_T^* is computed. Next, the algorithm moves back one stage and selects only the control variables that ensure the optimal end state x_T^*. Using the recursive relation given by Eq. (5.4), the set of states relevant for optimal policy are calculated by substituting the optimal state values into Eq. (5.4). As each element of this set is also an element of the state vector $\boldsymbol{X_{T-1}}$, interpolations during the calculation process can be avoided, yielding a reduction of the computing time. Thereafter, at stage $T-1$ the state with the minimum generated energy loss is determined based on the cost-to-go vector $\boldsymbol{G_k}$. The optimal control variable u_{T-1}^* at stage $T-1$ is the number of cells for that the energy loss is minimum. The algorithm moves back one stage at a time and calculates in the same way the next optimal state and thereafter the optimal control variable. As result, the optimal control policy is completely determined after the first stage $k = 0$ is reached. Algorithm 5.1 summarizes this calculation procedure.

Algorithm 5.1 Pseudocode for optimal policy calculation

1: **initialize:** $x_T^* \leftarrow \underset{x_T^i \in X_T}{\text{argmax}}\, J(x_T^i, u_{T-1}^i)$
2: **for** k = T-1 to 0 **do**
3: $\quad \overline{U_k} \leftarrow$ Set of the controls that ensure the optimal end state x_T^*
4: $\quad \overline{X_k} = x_{k+1}^* - f(x_{k+1}^*, \overline{U_k})$;
5: $\quad x_k^* = \underset{x_k^i \in \overline{X_k}}{\text{argmin}}\, G_k(x_k^i)$;
6: $\quad g_k^* = G_{k+1}(x_{k+1}^*) - G_k(x_k^*)$;
7: $\quad u_k^* = \underset{u_k^i \in \overline{U_k}}{\text{argmin}}\, g_k^*(u_k^i)$;
8: **end for**

5.5 Efficiency Enhancement

In this section different simulation studies are performed in order to benchmark the efficiency enhancement due to the reconfigurable architecture and proposed optimal control strategy.

5.5.1 Simulation setup

Two battery models with distinct topologies are built based on the same electrical model of a lithium-ion cell. More details about the electrical cell model can be found in Chen and Rincón-Mora (2006), among others. The reconfigurable battery model includes a supervisor control that determines the load of each cell according to the implemented strategy. For the sake of comparability, the cell capacities and internal resistances are consistent for both models. Note that the contact resistance within a battery are not considered in the model.

To investigate the performance of the proposed optimal control strategy over the battery's lifetime, a battery pack at different State of Health (SoH) levels, i.e., Beginning of Life (BoL), 75% SoH, 50% SoH, 25% SoH and 5% SoH, is adopted as an evaluation framework for the case study. The capacities and internal resistances are identified by measuring the lithium-ion cells of a battery pack at the considered SoH levels, before being put into the battery models.

Figure 5.5 shows the evolution of the cell capacities within the battery pack over its lifetime.

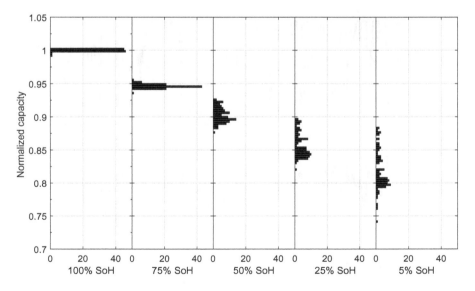

Fig. 5.5 Cell capacity distribution of the battery pack at different SoHs. The cell capacities are normalized to the nominal capacity.

5.5.2 Results

For each considered SoHs, the control strategy calculates the optimal control policy $\pi^* = \{u_0^*, \ldots, u_{T-1}^*\}$. Figure 5.6 illustrates the optimization results as the battery pack is at BoL.

It can be seen that more cells are deactivated during low power phases compared to high power phases. This is due to the fact that the power loss exhibits a dependence on the square of the current. Based on the power distribution created from the driving cycle and the optimal control policy, a lookup table is built and put into the supervisor control of the reconfigurable battery. The discharge process of the battery pack at each SoH is simulated along the driving cycle UDDS describing an urbane route. Thereafter, a comparison of energy efficiency of both topologies is carried out as shown in Fig. 5.7.

As can be seen from Fig. 5.7, a reconfigurable battery pack at BoL exhibits more than 98% energy efficiency, while less energy can be extracted from a non-configurable battery pack, leading to a 1.5% efficiency decrease. The same can be observed for cell packs with 75% and 50% SoH. Consequently, the efficiency enhancement due to the novel architecture is by 1.5% in the course of the first half of the battery's life.

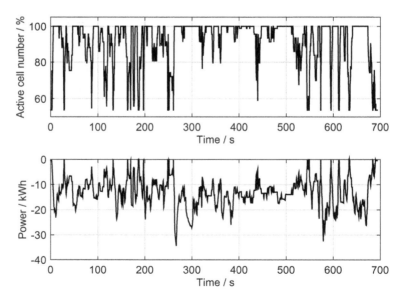

Fig. 5.6 Optimization results for the discharge phase of the UDDS cycle. The optimal control policy: number of active cells (normalized to the total number of battery cells) during the operating time (*top*) and battery power request (*bottom*).

Fundamental Aspects of Reconfigurable Batteries

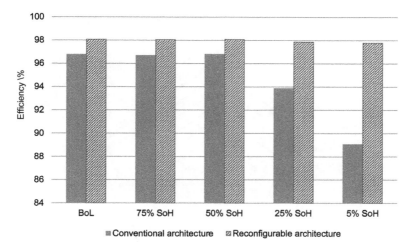

Fig. 5.7 Efficiency comparison between the conventional storage system and the proposed storage system.

However, over the second half, i.e., from 50% SoH until the end of its life, it can be observed that the lower the SoH of the battery pack, the worse the reduction of the energy efficiency of the conventional architecture. The simulation results reveal that when the SoH of the considered battery pack is 5%, the energy efficiency drops below 89%. In contrast, the reconfigurable architecture prevents a drop in the battery efficiency, which remains nearly 98% regardless the SoH of the battery pack. Hence, an energy efficiency enhancement of up to 9% is reached. Note that the marginal decrease of the proposed system efficiency is traced back to the significant internal resistance increase of aged cells.

The performance improvement by the reconfigurable architecture is explained by the fact that the efficiency of the conventional battery in the course of its lifetime is strongly affected by the weak cell within the battery pack as it determines the discharge time. Namely, the cell capacity spread is more significant within an aged battery pack, leading to more electrical energy remaining unusable in the battery. The proposed optimal control strategy instead calculates the optimal control policy for the actual cell capacities that will maximize the battery's efficiency.

To further investigate the performance of the proposed strategy, a comparison with a rule-based strategy is conducted. This strategy consists of deactivating the cell with the worst SoC during battery operation [Manenti et al. (2011)]. For comparison, the efficiency corresponding to each strategy

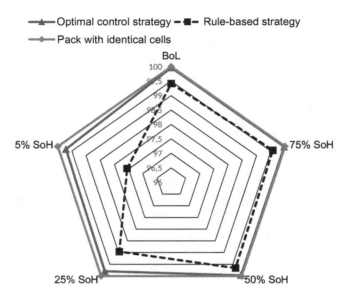

Fig. 5.8 Comparison between a battery pack implementing a reconfigurable topology with different control strategies and the ideal case where all cells are identical.

is normalized by the one from the ideal battery pack, i.e., the pack comprises identical cells. The result is illustrated in Fig. 5.8

It is clear that the proposed control strategy allows the battery pack to be operated almost at the ideal efficiency line during its lifetime, because it determines the optimal control policy for the given cell capacities. Instead, this is not possible with the rule-based strategy. Its main drawback is that the number of active cells cannot be adapted to the actual cell spread. In a used battery with low SoH, the cell spread is so high that more cells shall be deactivated to maximize the efficiency.

As a conclusion, in contrast to the rule-based strategy, the performance of the proposed optimal control is not affected by the SoH degradation, thus the optimal efficiency is ensured over lifetime.

5.6 Lifetime Enhancement

In the previous section the performance of reconfigurable architecture with the optimal control strategy compared to a non-configurable architecture is demonstrated by simulating the discharge behavior of a battery pack at different SoH values. This section seeks to assess how reconfigurable architecture enhances the battery lifetime.

5.6.1 Aging model

In the literature several aging studies for lithium-ion cells are carried out [Wang et al. (2014); Pinson and Bazant (2013); Käbitz et al. (2013)]. The cells are tested under conditions given in a predefined test matrix. During these accelerated aging tests, cells are characterized in fixed time steps by measuring their capacity and internal resistance values. The experimental data obtained are fitted using mathematical correlations that describe the dependency of the calender aging and cycle aging parameters on the aging factors such as temperature, voltage and depth of discharge. The identified aging parameters are used in aging model equations in order to quantitatively calculate the capacity fade and resistance rise of each cell in the battery for all conditions. Note that the inhomogeneity of the battery cells regarding the aging behavior is described using the normal distribution as was suggested in Paul et al. (2013).

5.6.2 Results

The aging of both battery systems was simulated using several successively repeated load profile. For both models the initial parameters, e.g., nominal capacity and internal resistance, correspond with those from the battery pack at its BoL, as shown in Fig. 5.5. Using the power request as an input the battery model calculates the response of each cell regarding voltage, SoC and temperature. The aging model later calculates the capacity loss and resistance rise for each cell during the time step Δt using the fitted mathematical equations. By taking into consideration the aging characteristic of the lithium-ion cells, the end-of-life criteria for the battery in the present lifetime study is set to 80%. Once the lifetime simulation is conducted, a comparison of the capacity spread within both storage systems over their lifetime is carried out, as shown in Fig. 5.9.

It can be observed that the cell spread within the conventional battery is more significant than in the case of a reconfigurable battery, namely over its lifetime. This improvement shown by the reconfigurable battery is due to the fact that the control strategy activates the cells with more energy content more often than those with less. Compared to cells with higher capacity, the weak cells are less stressed (lower temperature and fewer current peaks) and thus age more slowly. Moreover, this adapted aging inhomogeneity prevents these cells from damage. Consequently, the degradation rates of cell capacities within reconfigurable batteries are controlled in an appropriate manner to obtain a more even cell spread, to prevent weak cells from damage

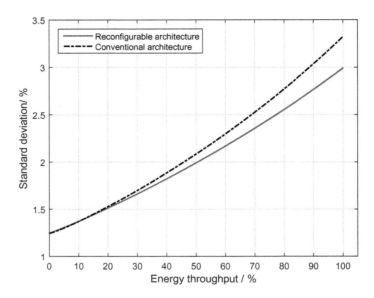

Fig. 5.9 Comparison of the cell capacity spread for the reconfigurable battery and conventional battery, assuming the same aging conditions (i.e., environment temperature and load profile) [Bouchhima *et al.* (2018)].

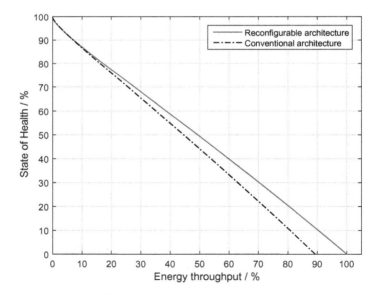

Fig. 5.10 Comparison of both storage systems' SoH over their lifetime [Bouchhima *et al.* (2018)].

and to enhance the SoH of the whole battery. Figure 5.10 illustrates the SoH of both storage systems with distinct topologies as a function of the battery energy throughput.

It can be seen that in the course of time the reconfigurable battery exhibits a significant better SoH than that of a conventional battery. While the capacity of the conventional storage system is restricted by the cell which shows the highest capacity degradation rate, capacity spread rise within the reconfigurable battery is reduced over its lifetime, thus improving the usable capacity of the whole system. In conclusion, the reconfigurable architecture extends the storage system's lifetime and leads to the fact that more energy can be provided until it fulfills its end-of-life criteria.

5.7 Summary

Reconfigurable architecture represents a potential solution against cell inhomogeneity within multi-cell batteries. After the advent of balancing circuits and their successful application on conventional battery architecture, the need for more system reliability leads to the necessity of more efficient solutions. The core of the reconfigurable architecture are the cell control unit and the energy management strategy implemented in the battery supervisor control. Optimizing the latter, the energy efficiency of the storage system can be maximized. To this end, the optimization theory was applied to formulate the energy distribution as an optimal control. Thereafter, based on dynamic programming and code vectorization, an optimization algorithm was established that solved the optimization problem by generating the optimal control policy for a known driving cycle. By implementing the proposed optimal control strategy, the energy efficiency gained due to the hardware architecture of reconfigurable batteries was much further enhanced. Simulations showed a significant improvement in energy efficiency over that of a non-reconfigurable storage system. In contrast to the rule-based strategy, the proposed control strategy allows the reconfigurable battery to be operated almost along the energy efficiency line of an ideal storage, where battery cells are identical over their lifetime. More benefits can be achieved with respect to a system's lifetime. To outline the impact of dynamic cell switching on the capacity degradation of the entire system, a semi-empirical aging model was used to reproduce the characteristic aging behavior observed of lithium-ion cells by incorporating the aging spread. Aging simulation results illustrated that contrary to conventional storage systems, cell spread increase within any reconfigurable battery was reduced with age. As a result a significant improvement concerning battery lifetime was reached. This result opens new perspectives for new cell classification processes in order to reduce costs. Moreover, the decrease of cell inhomogeneity could be a favorite condition for a second life use.

Bibliography

Bard, A. J. and Faulkner, L. R. (2001). *Electrochemical Methods: Fundamentals and Applications*, 2nd Edition (Wiley, New York).

Bouchhima, N., Gossen, M., Schulte, S., and Birke, K. P. (2018). Lifetime of self-reconfigurable batteries compared with conventional batteries, *J. Energy Storage* **15**, pp. 400–407.

Chen, M. and Rincón-Mora, G. A. (2006). Accurate electrical battery model capable of predicting runtime and 1–5 performance, *IEEE Trans. Energy Conver.* **21**, pp. 504–511.

Dantas De Melo, J., Calvet, J., and Garcia, J. (1990). Vectorization and multitasking of dynamic programming in control: Experiments on a CRAY-2, *Parallel Comput.*, pp. 261–269.

Daowd, M., Antoine, M., Omar, N., van den Bossche, P., and van Mierlo, J. (2013). Single switched capacitor battery balancing system enhancements, *Energies* **6**, pp. 2149–2174.

Dincer, I. and Rosen, M. A. (2007). *EXERGY: Energy, Environment and Sustainable Development* (Elsevier Ltd).

Elbert, P., Ebbesen, S., and Guzzella, L. (2013). Implementation of dynamic programming for n-dimensional optimal control problems with final state constraints, *IEEE Trans. Contr. Syst. Technol.* **21**, pp. 924–931.

Gallardo-Lozano, J., Romero-Cadaval, E., Milanes-Montero, M. I., and Guerrero-Martinez, M. A. (2014). Battery equalization active methods, *J. Power Sources* **246**, pp. 934–949.

Käbitz, S., Gerschler, J. B., Ecker, M., Yurdagel, Y., Emmermacher, B., André, D., Mitsch, T., and Sauer, D. U. (2013). Cycle and calendar life study of a graphite Li(NiMnCo)O2 Li-ion high-energy system. Part a: Full cell characterization, *J. Power Sources* **239**, pp. 572–583.

Kim, T., Qiao, W., and Qu, L. (2011). Series-connected self-reconfigurable multicell battery, *IEEE Appl. Power Electron. Conf. Expo. (APEC)* (Piscataway, NJ).

Kim, T., Qiao, W., and Qu, L. (2013). Real-time State of Charge and electrical impedance estimation for lithium-ion batteries based on a hybrid battery model, *28th Ann. IEEE Appl. Power Electron. Conf. Expo.* (Piscataway, NJ).

Lehner, S., Baumhöfer, T., and Dirk, U. S. (2016). Disparity in initial and lifetime parameters of lithium-ion cells, *IET Electr. Syst. Transport.* **6**, pp. 34–40.

Lu, L., Han, X., Li, J., Hua, J., and Ouyang, M. (2013). A review on the key issues for lithium-ion battery management in electric vehicles, *J. Power Sources* **226**, pp. 272–288.

Manenti, A., Abba, A., Merati, A., Savaresi, S. M., and Geraci, A. (2011). A new BMS architecture based on cell redundancy, *IEEE Trans. Ind. Electron.* **58**, pp. 4314–4322.

Paul, S., Diegelmann, C., Kabza, H., and Tillmetz, W. (2013). Analysis of aging inhomogeneities in lithium-ion battery systems, *J. Power Sources* **239**, pp. 642–650.

Pinson, M. B. and Bazant, M. Z. (2013). Theory of SEI formation in rechargeable batteries: Capacity fade, accelerated aging and lifetime predict, *The Electrochemical Society Meeting* (Toronto, Canada).

Srinivasan, V. and Wang, C. (2003). Analysis of electrochemical and thermal behavior of Li-ion cells, *J. Electrochem. Soc.* **150**, pp. 98–106.

von Stryk, O. and Bulirsch, R. (1992). Direct and indirect methods for trajectory optimization, *Ann. Oper. Res.* **37**, pp. 357–373.

Wang, J., Purewal, J., Liu, P., Hicks-Garner, J., Soukiazian, S., Sherman, E., Sorenson, A., Vu, L., Tataria, H., and Verbrugge, M. W. (2014). Degradation of lithium-ion batteries employing graphite negatives and nickel-cobalt-manganese oxide + spinel manganese oxide positives: Part 1, aging mechanisms and life estimation, *J. Power Sources* **269**, pp. 937–948.

Chapter 6

Volume Strain in Lithium Batteries

Jan Patrick Singer and Kai Peter Birke

6.1 Introduction

In the last few years, there has been much research activity on volume strain in lithium-based battery cells. We know that from the basic principle of reversible charge transfer between two electrodes, a change in electrode materials occurs during cycling. Depending on the materials and the resulting insertion/extraction reaction of the charge carrier with the host material, structural change is often followed by expansion, then contraction of the host. This results in mechanical stress and accelerated cyclic aging of a cell. A lot of research in the last decade was done to prevent volume changes by adding nanomaterials like carbon nano-tubes or special surface treatments. On a systems level, the cell stack is usually clamped with a defined pressure to decelerate aging processes. For future cell technologies, alloying and conversion electrodes are discussed. This material shows a significantly higher gravimetric as well as volumetric energy density. Naturally, some of them run through a dramatic volume change of several hundred percent. These expansions require the following: on the materials level, new surface technologies, additives, binder and electrolytes, and on the system level, not being able to avoid managing and tracking mechanical forces. For the understanding of expansion/contraction processes and their influence on cell and systems parameters, different measurement techniques are necessary. While volume strain is of a multi-scale investigation, it is of interest for material scientists, cell designers and systems engineers.

6.2 Fundamentals of Volume Strain

The functional principle of lithium-ion batteries is based on a reversible intercalation (insertion/extraction) reaction. Depending on the electrode

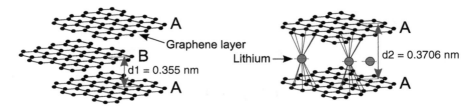

Fig. 6.1 Expansion of an *AB* layer-stacked graphite during lithium insertion. The bond length rises and results in a volume change of $\Delta V/V_0 = 10\%$.

materials, a widespread variety of chemicals reactions, driven by electromotive force occurs during intercalation. For example, during the intercalation of lithium (Li) into the anode material graphite (C_6) with the reaction

$$Li^+ + xe^- + C_6 \rightleftarrows Li_{x(0<x<1)}C_6, \tag{6.1}$$

the volume of the lattice expands about $\Delta V/V_0 \approx 10\%$ when the graphite is fully charged with lithium.

Figure 6.1 shows a simplified lithium intercalation into an *AB* layer stacking sequence. The distance between the layers $d = 0.335$ nm. After a fully lithiation of $C_6 \rightarrow LiC_6$ (Stage 1), the bond length rises to $d_2 = 0.3706$ nm. The expansion of the bond length is equivalent to $\Delta V/V_0$:

$$\frac{d_1}{d_2} = \frac{0.3706 \text{ nm}}{0.335 \text{ nm}} = 10.6\%, \tag{6.2}$$

Upon knowing the percentage volume change $\Delta V/V_0$, a simple calculation for the absolute volume change is possible. The amount of lithium n_{Li} for $C = 1$Ah of charge transfer into a graphite anode is:

$$n_{Li}(1\text{Ah}) = \frac{C}{eN_A} = 37.31 \text{ mmol}, \tag{6.3}$$

with the elementary charge e and the Avogadro constant N_A.

Applying the volume change $\Delta V/V_0 = 10.6\%$ as a coefficient of expansion α_{LiC6} for the changing from C to LiC_6 and the amount of lithium n_{Li}, together with the molar volume of carbon $\Delta V_{m,C}$, will calculate the change of the carbon volume due to lithium intercalation:

$$\Delta V_C(1\text{Ah}) = \alpha_{LiC6} \cdot n_{Li}(1\text{Ah}) \cdot 6 \cdot \Delta V_{m,C} = 0.13 \text{ cm}^3. \tag{6.4}$$

The molar volume of carbon in a hexagonal structure is $\Delta V_{m,C} = 5.31 \times 10^{-6} \text{ m}^3\text{mol}^{-1}$ [Bitzer and Gruhle (2014)].

In summary, the insertion/exertion of lithium into a host material causes a change in bond length, thus also resulting in a change in volume. Therefore, the type of insertion/exertion reaction that takes place does not matter, as it strongly depends on its type of materials. In literature, there are different classifications for the respective intercalation/insertion reactions. The term intercalation is often used for the storage of Li-ion in a layered host matrix that does not change the structure of the host during the intercalation reaction. Insertion is often used for alloying and conversion reactions, where a structural change, due to the Li insertion, in a metal matrix occurs. Since both terms — intercalation and insertion — cannot be differentiated and can be used interchangeably, we use the historical convention mentioned above [Winter et al. (1998)].

6.2.1 *Intercalation*

Intercalation is a reversible insertion/extraction of a species into a host structure. The term intercalation is referred to as a tunnel-like framework. Typical tunnel-like frameworks of crystal structures are one-dimensional (1D), two-dimensional (2D) and three-dimensional (3D) [Julien et al. (2016)].

Figure 6.2(a) shows the olivine (1D), layered (2D) and spinel (3D) insertion compounds which are associated with the dimensionality of the Li^+-ions transport.

During the intercalation of a species A (ion, atom or molecule) into a solid host network $\langle H \rangle$,

$$\langle H \rangle + xA \leftrightarrow A_x \langle H \rangle, \tag{6.5}$$

the crystal structure of the host compound is not changed or destroyed. A good cycle lifetime for the electrode results, it thereby expands the network

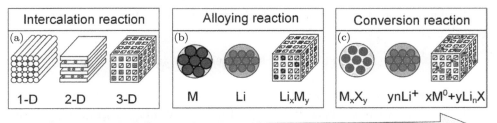

Fig. 6.2 Possible lithium insertion reaction as classification of (a) an intercalation reaction, (b) an alloying reaction and (c) a conversion reaction.

and the hence volume of the host. The dimension of the lithium transport and expansion of the host depends on the host materials. Materials with an intercalation reaction suffer from their low capacity due to low oxidation states. The resulting charge transfer is mostly limited to $1e^-/M$ [Palacin (2009)].

However, a phase change of the host might occur during the intercalation of lithium. This happens, for example, in LFP cathodes or LTO anodes. In literature both are attributed to intercalation reactions. There is no clear separation between the intercalation, alloying and conversion reaction.

6.2.2 Alloying

Alloying materials are under investigation, especially as anodes for the next-generation lithium-ion batteries. The advantages are high-energy densities, low costs, environmental friendliness and safe operation conditions [Zhang (2011)]. During the electrochemical alloying formation between lithium and a metallic (or semi-metallic) element, the materials will suffer a large volume of expansion/contraction [Julien et al. (2016)]. This prevents them from being practically utilized in commercial secondary batteries. During the reaction of tin with lithium, the number of atoms in every particle increases 440% [Palacin (2009)], which results in a very poor cycling lifetime due to tremendous volume expansions.

During an addition reaction, the parent phase of M runs through a phase change to LiM_x. Typically, this occurs during the lithiation of crystalline Si, Sn, Al and Sb while they change their crystal structure [Zhang (2011)].

$$Li + xM \leftrightarrow LiM_x. \tag{6.6}$$

A second group for alloying reaction are the displacement reactions. Lithium reacts with one component M of an alloying compound MN_y. A second component N, that might be active/inactive towards lithium, is displaced from the former phase. A general expression of a replacement reaction is:

$$Li + xMN_y \rightarrow LiM_x + xyN. \tag{6.7}$$

The resulting cracks and pulverization can be reduced by replacing bulk material with nanostructured particles. Commercial graphite anodes contain amorphous carbon with embedded Sn-Co nanoparticles. The volume expansion of the Li-Sn alloying reaction is buffered by the matrix of the carbon. Additional, the cobalt prevents the crystalline phase transition [Palacin (2009)].

6.2.3 Conversion

Conversion electrodes have been established as primary cells and are still under production today. For the establishment of conversion materials as secondary cells, reversibility is the most challenging. Especially for high voltage cathodes, conversion materials are interesting candidates. Conversion electrodes using binary transition-metals $M_y X_Z$ to benefit from the advantage of nanostructured materials in a reaction pathway change. The metal compound M can be either iron, manganese, nickel, cobalt or copper. They can react either with an oxide (x=O), sulfide (x=S), flouride, (x=F), nitride (x=N), phosphite (x=P) and hydrogen (x=H) [Cabana et al. (2010)] with the conversion reaction

$$M_x X_y + yne^- + ynLi^+ \leftrightarrow xM^0 + yLi_nX, \qquad (6.8)$$

while n is the formal oxidation state of the anion. Thereby, the morphology of the binary compounds M_X changes. The metal M^0 decomposes under a complete reduction to nanoparticles of 5–12 nm diameter during the formation of a lithium binary compound Li_nX. The 3D metal nanoparticles of M^0 are finally embedded in a Li_nX matrix [Julien et al. (2016)]. Due to the high oxidation states of phases during the conversion reaction, 2 to 6 electrons can be transferred. A high-energy density results [Armand and Tarascon (2008); Cabana et al. (2010)]. Unfortunately, there is a general trend between oxidation states (high-energy density) and the volume change during the conversion reaction. A higher oxidation state means that a large amount of Li^+ reacts per mol of transition-metals $M_y X_Z$ and results in volume changes of up to several hundred per cent [Ponrouch et al. (2012)].

6.3 Volume Strain on Cells Level

On the cell level, the cyclic volume strain causes a variety of different effects. Mechanical consequences of the insertion/extraction reactions are particle cracks, compared with Fig. 6.3. The internal cell pressure increases and transfers the mechanical stress on the particles. Frequently, electrode materials are porous and the particle cracks increase until a piece of the particle breaks. Since this piece is not in contact with the active material anymore, it can be considered as a loss capacity. Furthermore, the active material particles often present a low conductivity inbetween and to the current collector. To enhance the conductivity, a conductive carbon black is added in the electrode slurry, see Fig. 6.3. Due to the expansion, the conductivity paths break up and increase the internal resistance and an

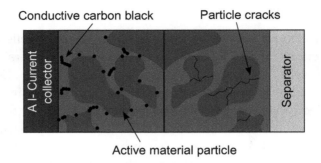

Fig. 6.3 Mechanical effects of volume changes on the cell level. Losing the conductivity of carbon black and the particle cracks. As a consequence the cyclic aging is accelerated.

accelerated aging behavior follows. To avoid this, particle cracks and broken-up conductivity paths, cell manufacturers add a binder to the electrodes. The binder absorbs the mechanical impact of the insertion/extraction process. Unfortunately, the binder material (e.g., Polyvinylidene flouride, PVDF) ages as well and lowers the absorbing properties. An aged binder accelerates the cracking and breaking up of conductivity paths.

Aging effects on the cells level can also result in an increase of volume strain over its lifetime. On both electrodes, surface films are built up during cycling. On the cathode, there is a very small conductive interface, and on the anode side, the often discussed and investigated solid electrolyte interface (SEI). This SEI is at least more than 10 times thicker than the conductive interface. It is formed during the first cycle and growth by the loss of liquid electrolyte and active material during its cyclic lifetime. Increasing surface films mean less free space for the expansion and contraction process. The internal cell pressures increase and accelerate the fraction of particles. Not only do particles crack, so too do the important surface films. Under the loss of electrolyte and active material, stress-induced cracks on surface films are healed. Another effect of rising cell pressure happens to the separator. Fine pores of the separator, which are soaked by electrolyte enable the diffusion of Li-ions. If the pressure is too high, the diffusion process will be blocked by sealed pores.

All these effects and consequences happen only on the materials level and can only be avoided by consequent progression of every part of a cell.

6.4 Volume Strain on Systems Level

Usually, a battery system manufacturer purchases an amount of cells at a cell producer. Based on that, the cell is like a black box. Consequently, there is no influence on the materials level for this composition. However, systems designers have come up with a number of suggestions on how to extend the systems level's lifetime, with regard to volume changes. First of all, the cell needs to be run within the recommended specifications. In reality, this condition cannot always be satisfied, because at low temperatures, high C-Rates and low SoC, there is a possibility of lithium plating. Electrochemical deposition of metallic lithium occurs when the half-cell potential of graphite drops below 0 V versus Li/Li$^+$. Usually, the operator cannot monitor half-cell potential, respectively local potentials of the electrodes, due to inhomogeneous intercalation. Consequently, the cell might run into a plating condition. Metallic lithium plating is an additional increase of the cells' volume. Consequently, deposited lithium cannot be intercalated. A theoretical relation between volume expansion $V_{\exp}(C_{\mathrm{pl}})$ and the amount of plated lithium C_{pl} is according to Bitzer and Gruhle (2014):

$$V_{\exp}(C_{\mathrm{pl}}) = \frac{C_{\mathrm{pl}}}{e \cdot N_{\mathrm{A}}}(V_{\mathrm{m,Li}} - \alpha_{\mathrm{LiC6}} \cdot 6 \cdot V_{\mathrm{m,C}}). \tag{6.9}$$

Not only might low temperatures and low SoC cause volume expansion, high cell temperatures, also as a function of the C-Rate and high SoC might start a decomposition of the electrolyte. For both voltage and temperature, every electrolyte has an operation window. If the operation condition runs out of the stability window of the electrolyte, gassing can occur [Michalak et al.]. During different gassing phases, volume expansion of up to several hundred percent are possible [Self et al. (2015)]. This strongly depends on the electrolyte composition. Another source for the evolution of gas is due to the formation of new SEI layers. During the formation of a new cell, the SEI is built as a protection layer. Therefore, small amounts of electrolyte are decomposed under the evolution of gas. After the formation, the gas will be released and the cells are evacuated. As discussed before, due to particle cracks, new SEI layers are formed. But this time, no evacuation takes place and the gas is located within the cell housing. Consequently, the formation gas enhances the cells' pressure.

A mechanical design's aspect on the systems level is the compression of a cell array. A row of cells within a module is often fixed with compression bars and end plates. Between the cells, spacers provide a means for compression [Mohan et al. (2014)]. With the help of the end plates and the compression bar, an initial pressure is applied to the cells. A constant external pressure means an increasing cell pressure, when the volume change increases due to an aging mechanism. If the external cells' pressure gets too high, the positive effects are negligible and negative effects (like accelerated aging) predominate. At present, pressure tracking systems are under research in the science and industrial industries. Several methods can be envisaged:

- Mechatronic management
- Mechanical tracking
- Intelligent materials

Mechatronic management tracks cell pressure by a servomotor. A force sensor (compare with Sec. 6.5) detects the current pressure and sends the signal to superordinate management systems. This can be included within a battery management system (BMS). This management system controls the servomotor and ensures an optimized cell pressure during the lifetime of the battery pack. Control systems based on electrical and electronic parts can react very fast, and are reliable and accurate. Their disadvantage is the extra complexity due to additional sensors, actuators and the control system itself. Also, to the detriment of the energy density, the components require additional space within a battery module. The mechanical system works on the principle of a preloaded spring, which ensures that external force is provided to the cell rack and it absorbs the volume strain-induced force change. No additional sensors and control systems are required for this method and the system's integration takes place easily due to compact mounting position of the spring. However, the spring cannot re-stress the cell and ages mechanically. This results in a non optimal cell pressure and might accelerate aging effects. Currently under trial are spring systems that age at the same rate as the volume increase of a cell. With such a system, a constant cell pressure is now possible. On the cell's level, intelligent materials might be beneficial in handling the challenges of volume changes. The binder and the separator, as well as the housing of the cell and spacers between the cells, absorb some force. Intelligent materials can change their absorbent properties under defined conditions. Also, special additives are currently being discussed and researched.

6.5 Measurement Techniques

There are different possibilities for measuring the volume strain of a cell. In fact, all measurement techniques are based on an indirect principle. Since we have different types of housing for the electrodes, not all measurement

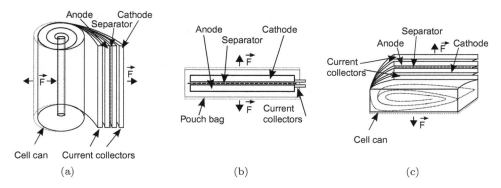

Fig. 6.4 Different cell formats of lithium-ion cells: (a) round, (b) pouch and (c) prismatic. Depending on the electrode stacking and stiffness of the housing, the cells' internal force F acts in different directions.

techniques are suitable for every type of housing. Below are three types of housing (see Fig. 6.4):

- Softcase (Pouch cells)
- Hardcase (Round cells, prismatic cells, coin (button) cells)
- Test cells (Half-cells)

The housing of the pouch cell, often called "coffee bag" is made of an aluminum laminated film. Under a vacuum atmosphere, the electrode batch is sealed. Because of the vacuum atmosphere inside the pouch bag and its thickness of up to around 150 μm the housing expands during lithiation. This enables a wide range of different measurement techniques. The internal force F impacts most of the separator and the housing in one direction. However, it is more difficult to measure volume changes in hardcases. Compared to the flexible aluminium laminated foil of the pouch cell, the hardcase is often made from stainless steel mixed with nickel for extra hardness. There are two types of hardcases that are common in secondary lithium-ion cell technologies. In both, the round cell (in format 18650 and often called the laptop cell) and the prismatic cell, a "jelly roll" electrode configuration, are packaged into a metal case. For a round cell in format 18650, the steel can is about 140 μm thick [Hagen et al. (2015)], the thickness of the prismatic cell is bigger at around 3.6 mm [Hatchard et al. (2001)]. Because of the thick and stainless steel case and the kind of winding of the electrode stacks, the cell does not expand at the same rate as the pouch bag. In addition, a battery mandrel is used to fix the rolled electrodes of the prismatic and round cells. The hollow mandrel is made of nylon and absorbs a part of the force that acts to the middle of the cell.

The third type of housing — test cells — are often mentioned for the characterization of new material composition. Often, a small sample of a new material is synthesized in a coin format and tested in a special test cell that allows the sample to run in a laboratory format. This is, inter alia, often used to test materials when the after surface treatments for the reduction or prevention of volume changes.

6.5.1 *Unpressurized*

Measuring volume change without applying an external force on the cell involves a multiscale investigation. Figure 6.5 shows several methods whose application depends on the type of measurement (single electrode versus cell) and dimensions. A common application is the electrochemical dilatometry on the materials level for test cells. During electrochemical lithiation, the single electrode cell, called the working electrode, is working against the counter electrode. A very sensitive displacement sensor measures the expansion during the cycling process. Often, a very small mechanical load to ensure the conductivity of the electrode is applied, together with the self-weight of the sensor [Rieger *et al.* (2016)], resulting in a total load of about 1 N. The result of the electrochemical dilatometry is a one-dimensional displacement in μm on a single spot of the electrode. Possible inhomogeneities due to production faults are not identifiable.

On the cells' level, the overall one-dimensional thickness change can also be measured by a displacement sensor. Note that a single electrode material may have a negative volume strain during lithiation, see Table 6.1, and the total thickness of a cell may differ from that of the single electrode. Additionally, the separator and the housing of the cell absorb some strain. Here, it is possible to mount a sensor on the same spot in a X-Y direction: one for the button side and one for the upper side [Rieger *et al.* (2016)]. This setup enables the acquisition of displacement data in both the $-z$ and $+z$ directions. It is also possible to set up a number of sensors at different locations [Oh *et al.* (2014)]. Usually, the middle of the cell and its edges are of great interest because the strongest swelling takes place in the former, while the areas in the latter are influenced by the production quality. There are different displacement sensors. The most suitable for battery cell application are the LVDT (Linear Variable Differential Transformer) sensors, whose basic components consist of a movable core of magnetic material, and one primary and two secondary coils. The sensor works to the transformer principle: the output voltage $V_{\text{out}} = 0$, if there is no displacement D. Depending on the sensitivity of the sensor M, voltages V_A and V_B are induced

Table 6.1 Volume Strain $\Delta V/V_0$ and Potential versus Li/Li^+ of lithium-storage compounds with respect to the limiting composition based on [Woodford et al. (2012)].

Lithium storage compound	Limiting composition	Volume strain $\Delta V/V_0$ (%)	Insertion reaction	Potential vs. Li/Li^+ (V)	Reference
Cathode materials (Li-extraction)					
$LiCOO_2$ (LCO)	$Li_{0.5}COO_2$	+1.9	2D-intercalation	3.9	[Reimers and Dahn (1992)]
$LiNi_{0.8}Co_{0.2}O_2$	$Li_{0.5}Ni_{0.8}Co_{0.2}O_2$	−2.04	2D-intercalation	3.6	[Saadoune and Delmas (1998)]
$LiNi_{0.8}Co_{0.15}Al_{0.05}O_2$ (NCA)	$Li_{0.5}Co_{0.8}Co_{0.15}Al_{0.05}O_2$	−1.16	2D-intercalation	3.7	[Itou and Ukyo (2005)]
$LiNi_{1/3}Mn_{1/3}Co_{1/3}O_2$ (NMC)	$Li_{0.47}Ni_{1/3}Mn_{1/3}Co_{1/3}O_2$	2.44	2D-intercalation	3.7	[Yoon et al. (2006)]
$LiFePO_4$ (LFP)	$FePO_4$	−6.6	1D-intercalation	3.4	[Meethong et al. (2007)]
$LiMn_2O_4$ (LMO)	Mn_2O_4	−7.3	3D-intercalation	4.0	[Hunter (1981)]
$LiNi_{0.5}Mn_{1.5}PO_4$ (LNMO)	$Ni_{0.5}Mn_{1.5}PO_4$	−6.2	3D-intercalation	4.7	[Ariyoshi et al. (2004)]
S	Li_2S	80	Conversion	2.4	[Manthiram et al. (2014)]
Anode materials (Li-insertion)					
C_6 (graphite)	LiC_6	12.8	2D-intercalation	0.1	[Ohzuku et al. (1993)]
$Li_4Ti_5O_{12}$ (LTO)	$Li_7Ti_5O_{12}$	0	3D-intercalation	1.5	[Ohzuku et al. (1995)]
Si	$Li_{4.4}Si$	310	Alloying	0.3	[Beaulieu et al. (2001)]
Sn	$Li_{4.4}Sn$	255	Alloying	0.58	[Nitta et al. (2015)]
Li	Li_2O_2	30	Conversion	2.96	[Albertus et al. (2014)]

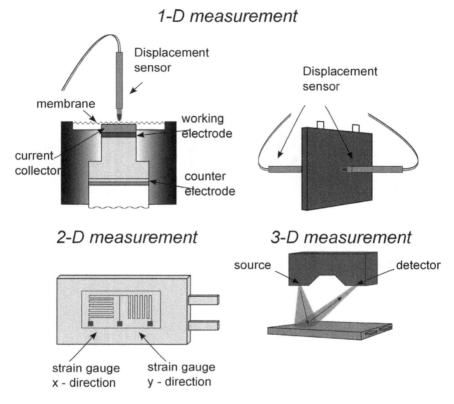

Fig. 6.5 Unpressurized measurement techniques for different dimensions of lithium-ion cells. One-dimensional measurements are possible by using a displacement sensor, while two-dimensional measurements use a strain gauge and an optical system allows three-dimensional measurements.

in the secondary coil. The displacement can be calculated as:

$$D = M \frac{V_A - V_B}{V_A + V_B}. \quad (6.10)$$

A suitable possibility for measuring a 2D surface strain is via two strain gauges that contain two single electrode areas placed at a 90° angle. One measures the x-direction strain; the other, the y-direction. Because of the strain on the resistive foil due to expansion, the electric conductance of the sensor array changes — it depends on the geometry of the conductive lines. Because of this design, the zigzag-patterned conductive lines enable the detection of a very small amount of stress ϵ due to the multiplication of the resistance change ΔR over the resistance of the un-deformed gauge R_G.

The gauge factor GF

$$GF = \frac{\Delta R/R_\text{G}}{\epsilon}, \qquad (6.11)$$

together with a Wheatstone bridge is able to detect the strain as a voltage signal v with an excitation voltage V_in:

$$v = \frac{V_\text{in} GF \epsilon}{4}. \qquad (6.12)$$

One possibility to measure the 3D scale of any thickness change is via an optical scan of the surface. For example, a structured-light scanning system can be used to scan the surface [Rieger et al. (2016)]. For stationary objects, like a battery cell, the structured-light is preferred for use as compared to laser line triangulation. The light source of the scanning system projects a light beam (changing line structures) onto the cell and measures the reflected signal via a detector. The system scans the whole surface of a cell and calculates via a software the expansion of every scanning point. To calculate the position, the sensor detects the slope and position of the reflected line structure. A software using a closed point algorithm generates a 3D image as a mesh of many points [Bell et al. (1999)].

Based on triangulation calculations, the height $Z_\text{H}(x,y)$ of the cell's surface can be calculated at any point in a x-y direction. Knowing the distance L between the detector and a reference, the distance d between the source, detector and period p of the modulated light-beam, $Z_\text{H}(x,y)$, can be evaluated by Sansori et al. (1994):

$$Z(x,y) = \frac{Lp\Delta\phi(x,y)}{2\pi d + p\Delta\phi(x,y)}. \qquad (6.13)$$

The phase information $\Delta\phi(x,y)$ between the phase before and after the expansion, with the lateral shift $\overline{BA}$ of ray due to surface conditions, can be written as:

$$\Delta\phi(x,y) = \frac{2\pi}{p}\overline{BA}. \qquad (6.14)$$

6.5.2 *Pressurized*

Another possibility for online diagnostic are pressurized measurements, see Fig. 6.6. Applying an external pressure p on the surface of a pouch or prismatic cell extended the cyclic life span [Cannarella and Arnold (2014b)]. There are several reasons for constraining lithium-ion cells, e.g., preventing delamination, ensuring conductivity (ionic and electric), avoiding cracks, etc.

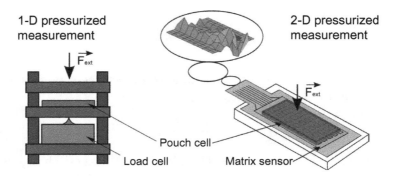

Fig. 6.6 Measurement techniques for pressurized one- and two-dimensional measurements of lithium-ion cells.

The battery designer needs to make sure not to apply too much pressure, otherwise the cycle stability decreases. During its lifetime, the internal cell pressure increases because of the aging mechanism, see Fig. 6.7(c), and it requires the external pressure to be readjusted. The total load of the cell can be assumed with a load cell. It measures the deformation of a strain gauge due to applied force. The measurement principle is comparable to the strain gauge in Sec. 6.5.1. Usually, the load cell is connected to a Wheatstone bridge and induced with a very small voltage drop of some Milivolts. For this reason, a high resolution analog/digital converter, typically 24-bit, is required. The load cells are available in different versions. If single cells are mounted in a battery pack and tightened with a threaded rod, a through-hole load can be integrated. In this case, the bulk force of the constrained battery pack can be monitored. There are also piezoelectric, hydraulic and pneumatic load cells available. However, these are less suitable for this kind of application.

For 2D pressure mapping, the application of a matrix tactile sensor system can be envisaged. This measurement principle works either on resistance or capacitance effects. A defined number of columns and rows are printed on the upper respective button side on a dielectric foil. They are placed at a right angle. Every intersection of a row and column represents a mapping point. While applying pressure on an intersection, the resistance or the capacitance (depending on the measurement principle) of the spot changes. Every row and column is connected to a monitoring system that measures via logic, for example, the voltage drop of a spot. A software is then able to calculate and visualize a pressure map of the surface. The sensor array can be placed between a pouch cell and a tension plate. Because of the

Table 6.2 Overview of different techniques for indirect volume change measurements and their possible application to different cell types.

Cell type	Displacement	Gauge strain	3D-scan	Load cell (press.)	2D-mapping (press.)
Cylindrical	-	x	x	-	-
Prismatic	x	x	x	x	-
Pouch	x	x	x	x	x
Testing	x	-	-	x	-

inhomogeneous volume strain of the cell, the mapping system allows one to observe this inhomogeneous strain during cycling.

Table 6.2 summarizes the previously discussed measurement techniques and recommends their application to different cell types.

6.6 State Diagnostics

For operating a lithium-ion cell in proper conditions, the knowledge about the actual state of the cell is very important. There are several states that describe the behavior of a cell. The two fundamental and most important states are the State of Charge (SoC) and State of Health (SoH). Unfortunately both are, depending on the cells chemistry, very challenging for estimation. For example, intercalation electrodes like NCM, NCA / C6 show an approximately linear increasing voltage-charge relationship which makes the SoC estimation easier, as seen in Fig. 6.7(a). LFP with a voltage plateau during the two-phase region where a very small voltage change result is more challenging and needs additional information for state diagnostics. This additional physical variable might be the cell's pressure P_{nom}. Currently the focus in science and research is on post lithium-ion batteries like lithium-air, lithium-sulfur, respectively, and metal-air in general. Unfortunately all of them undermine dramatic volume changes in the range of some alloying reactions. For handling this volume changes, but also to enlarge the lifetime of the cell chemistries, a mechatronic management for tracking a constant pressure is required. This property can be of benefit while the cell's pressure is measured. In addition, stress can be used to provide real time measurement data for SoC and SoH estimation [Cannarella and Arnold (2014a)].

6.6.1 *SoH diagnostics*

A major aging effect is the growth of surface layers like the Solid Electrolyte Interface (SEI), see Fig. 6.8, on the anode due to dissolution of the liquid

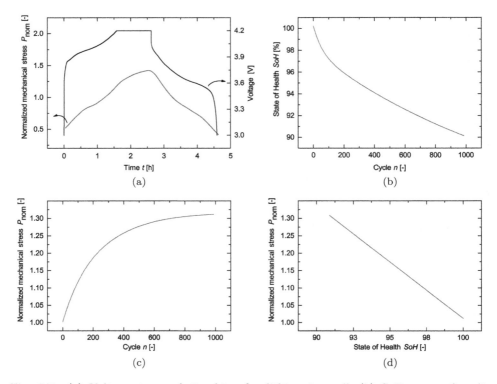

Fig. 6.7 (a) Voltage-stress relationship of a lithium-ion cell, (b) SoH curve of cyclic aging, (c) the relationship between the maximum of stress P_{nom} and (d) cyclic aging and the linearity between P_{nom} and SoH.

electrolyte as a result of volume expansion and contraction. This results in a loss of usable capacity, and hence a decreasing SoH. Figure 6.7(b) shows the decreasing cycling lifetime. In real application, the SoH diagnostic is quite challenging due to nonlinear aging effects. This also further depends on the actual SoH from the previous cycles. In Fig. 6.7(c), the increasing maximum cell pressure of an aging cell is illustrated. Note also that the stress curve depends on external influences. A linear relationship between the rise of the maximum stress during cyclic aging and the loss of capacity is found in [Cannarella and Arnold (2014a)] and plotted in Fig. 6.7(d). The application of a SoH algorithm might contain this information. Many different aging effects of a cell results in a total growth of the cell's pressure.

6.6.2 SoC diagnostics

As discussed earlier the SoC estimation is very challenging, but also very important because of safety issues. The stress-SoC relationship should be

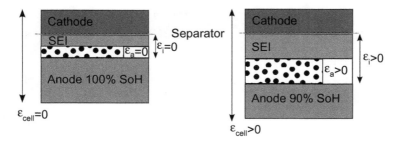

Fig. 6.8 Anodic strain growth due to the aging effects of increased solid electrolyte interface (SEI) layers and increased volume change of the active material.

used especially for chemistries with flat voltage characteristics. To measure the mechanical stress as a support function for SoC determination, it

- must be independent from temperature influences
- must be independent from I/R drops due to high C-Rates
- must be able to detect internal self-discharge (compared to coulomb counting, which measures only an external charge transfer)

When using the stress-SoC relationship, there should be a dependence on the stress-SoC curve and SoH, as seen in Fig. 6.7(c). Due to the aging, the irreversible thickness change results in both, a change of the slope and an increase of the maximum stress level in the stress-SoC curve.

Consequently, we can write for intercalated lithium q_a in the anode:

$$q_a = q_0(SoC)(SoH), \qquad (6.15)$$

with the initial cell capacity q_0. In the following section, only the expansion of the anode (= max. stress) at 100% SoC, according to [Cannarella and Arnold (2014a)], is considered. The constant modulus of elasticity E and the strain ϵ can be experimentally determined by the stress-strain curve at 100% SoC and 100% SoH, respectively. A typical stress-strain relationship is shown in Fig. 6.9.

The total strain ϵ_{cell} of the cell can be written as the sum of the irreversible strain because of aging ϵ_i and the increasing strain of the anode ϵ_a, which also depends on SoH:

$$\epsilon_{\text{cell}} = \epsilon_i + \epsilon_a. \qquad (6.16)$$

Here, ϵ_i is representative for the growth of surface films likes the SEI.

At 100% SoH, both ϵ_i and ϵ_a are defined as 0.

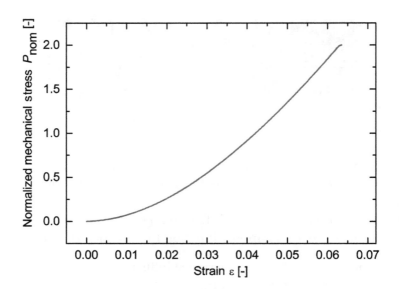

Fig. 6.9 Strain-stress relationship of a lithium-ion pouch cell.

During the lifetime of the cell, ϵ_i and ϵ_a increase proportionally with constants k_i and k_a, depending on SoH:

$$\epsilon_i = k_i(1 - SoH), \quad (6.17)$$

$$\epsilon_{ai} = -k_a(1 - SoH). \quad (6.18)$$

Combining the fundamental relationship of strain ϵ, σ and constant modulus of elasticity E with the initial stack pressure σ_0, the stress of the charged cell is

$$\sigma = \sigma_0 + E\sigma_{cell}, \quad (6.19)$$

respectively, replacing σ_{cell}

$$\sigma = \sigma_0 + E(k_i - k_a)(1 - SoH). \quad (6.20)$$

Bibliography

Albertus, P., Lohmann, T., and Christensen, J. (2014). Overview of LiO2 battery systems, with a focus on oxygen handling requirements and technologies, in N. Imanishi, A. C. Luntz, and P. Bruce (eds.), *The Lithium Air Battery: Fundamentals* (Springer New York), ISBN 978-1-4899-8062-5, pp. 291–310, doi:10.1007/978-1-4899-8062-5_11.

Ariyoshi, K., Iwakoshi, Y., and Nakayama, N. (2004). Topotactic two-phase reactions of Li Ni 1/2 Mn 3/2 O4 P 4332 in nonaqueous lithium cells, *J. Electrochem. Soc.* **151**, 1, pp. 296–303, doi:10.1149/1.1639162.

Armand, M. and Tarascon, J. (2008). Building better batteries, *Nature* **451**, February, pp. 685–657, doi:10.1038/451652a.

Beaulieu, L. Y., Eberman, K. W., Turner, R. L., and Krause, L. J. (2001). Colossal reversible volume changes in lithium alloys, *Electrochem. Solid St.* **C**, pp. 7–10, doi:10.1149/1.1388178.

Bell, T., Li, B., and Zhang, S. (1999). Structured light techniques and application, in *Wiley Encyclopedia of Electrical and Electronics Engineering* (John Wiley & Sons, Inc.), ISBN 9780471346081, doi:10.1002/047134608X.W8298, http://dx.doi.org/10.1002/047134608X.W8298.

Bitzer, B. and Gruhle, A. (2014). A new method for detecting lithium plating by measuring the cell thickness, *J. Power Sources* **262**, pp. 297–302, doi:10.1016/j.jpowsour.2014.03.142, http://dx.doi.org/10.1016/j.jpowsour.2014.03.142.

Cabana, B. J., Monconduit, L., Larcher, D., and Palacín, M. R. (2010). Beyond intercalation-based Li-ion batteries: The state of the art and challenges of electrode materials reacting through conversion reactions, *Adv. Energy Mater.* **22**, 35, pp. 170–192, doi:10.1002/adma.201000717.

Cannarella, J. and Arnold, C. B. (2014a). State of health and charge measurements in lithium-ion batteries using mechanical stress, *J. Power Sources* **269**, pp. 7–14, doi:10.1016/j.jpowsour.2014.07.003.

Cannarella, J. and Arnold, C. B. (2014b). Stress evolution and capacity fade in constrained lithium-ion pouch cells, *J. Power Sources* **245**, pp. 745–751, doi:10.1016/j.jpowsour.2013.06.165, http://dx.doi.org/10.1016/j.jpowsour.2013.06.165.

Hagen, M., Hanselmann, D., Ahlbrecht, K., Maça, R., Gerber, D., and Tübke, J. (2015). Lithium-sulfur cells: The gap between the state of the art and the requirements for high energy battery cells, *Adv. Energy Mater.* **5**, 16, doi:10.1002/aenm.201401986.

Hatchard, T. D., Macneil, D. D., Basu, A., and Dahn, J. R. (2001). Thermal model of cylindrical and prismatic lithium-ion cells, *J. Electrochemical Soc.* **148**, 7, pp. A755–A761, doi:10.1149/1.1377592.

Hunter, J. C. (1981). Preparation of a new crystal form of manganese dioxide: λ-MnO2, *J. Solid State Chem.* **39**, pp. 142–147.

Itou, Y. and Ukyo, Y. (2005). Performance of $LiNiCoO_2$ materials for advanced lithium-ion batteries, *J. Power Sources* **146**, August, pp. 39–44, doi:10.1016/j.jpowsour.2005.03.091.

Julien, C., Mauger, A., Vijh, A., and Zaghib, K. (2016). Principles of intercalation, in *Lithium Batteries: Science and Technology*, Chap. 3 (Springer, Cham), ISBN 978-3-319-19108-9, pp. 69–90, http://dx.doi.org/10.1007/978-3-319-19108-9.

Manthiram, A., Fu, Y., Chung, S.-H., Zu, C., and Su, Y.-S. (2014). Rechargeable lithium-sulfur batteries, *Chem. Rev.* **114**, 23, pp. 11751–11787, doi:10.1021/cr500062v.

Meethong, B. N., Huang, H.-Y., Speakman, S. A., Carter, W. C., and Chiang, Y.-M. (2007). Strain accommodation during phase transformations in olivine-based cathodes as a materials selection criterion for high-power rechargeable batteries, *Adv. Funct. Mater.* **17**, pp. 1115–1123, doi:10.1002/adfm.200600938.

Michalak, B., Sommer, H., Mannes, D., Kaestner, A., Brezesinski, T., and Janek, J. (2015). Gas evolution in operating lithium-ion batteries studied in situ by neutron imaging, *Sci. Rep.* **5**, pp. 1–9, doi:10.1038/srep15627, http://dx.doi.org/10.1038/srep15627.

Mohan, S., Kim, Y., Siegel, J. B., Samad, N. A., and Stefanopoulou, A. G. (2014). A phenomenological model of bulk force in a Li-ion battery pack and its application to State of Charge estimation, *J. Electrochem. Soc.* **161**, 14, pp. A2222–A2231, doi:10.1149/2.0841414jes, http://jes.ecsdl.org/cgi/doi/10.1149/2.0841414jes.

Nitta, N., Wu, F., Lee, J. T., and Yushin, G. (2015). Li-ion battery materials: Present and future, *Mater. Today* **18**, 5, pp. 252–264, doi:10.1016/j.mattod.2014.10.040, http://dx.doi.org/10.1016/j.mattod.2014.10.040.

Oh, K. Y., Siegel, J. B., Secondo, L., Kim, S. U., Samad, N. A., Qin, J., Anderson, D., Garikipati, K., Knobloch, A., Epureanu, B. I., Monroe, C. W., and Stefanopoulou, A. (2014). Rate dependence of swelling in lithium-ion cells, *J. Power Sources* **267**, pp. 197–202, doi:10.1016/j.jpowsour.2014.05.039, http://dx.doi.org/10.1016/j.jpowsour.2014.05.039.

Ohzuku, T., Iwakoshi, Y., and Sawai, K. (1993). Formation of lithium-graphite intercalation compounds in nonaqueous electrolytes and their application as a negative electrode for a lithium-ion (shuttlecock) cell, *J. Am. Chem. Soc.* **140**, 9.

Ohzuku, T., Ueda, A., and Yamamota, N. (1995). Zero-strain insertion material of Li[Li1/3Ti5/3]O4 for rechargeable lithium cells, *J. Electrochem. Soc.* **142**, 5.

Palacin, M. R. (2009). Recent advances in rechargeable battery materials: A chemist's perspective, *Chem. Soc. Rev.* **38**, 9, pp. 2565–2575, doi:10.1039/b820555h.

Ponrouch, A., Taberna, P.-l., Simon, P., and Palacín, M. R. (2012). On the origin of the extra capacity at low potential in materials for Li batteries reacting through conversion reaction, *Electrochim. Acta* **61**, pp. 13–18, doi:10.1016/j.electacta.2011.11.029, http://dx.doi.org/10.1016/j.electacta.2011.11.029.

Reimers, J. N. and Dahn, J. R. (1992). Electrochemical and in situ x-ray diffraction studies of lithium intercalation in LixCo02, *J. Electrochem. Soc.* **139**, 8, pp. 2091–2097, doi:10.1149/1.2221184, http://jes.ecsdl.org/content/139/8/2091.abstract.

Rieger, B., Schlueter, S., Erhard, S. V., Schmalz, J., Reinhart, G., and Jossen, A. (2016). Multi-scale investigation of thickness changes in a commercial pouch type lithium-ion battery, *J. Energy Storage* **6**, pp. 213–221, doi:10.1016/j.est.2016.01.006, http://dx.doi.org/10.1016/j.est.2016.01.006.

Saadoune, I. and Delmas, C. (1998). On the LixNi0.8Co0.2O2 system, *J. Solid State Chem.* **15**, 136, pp. 8–15.

Sansori, G., Biancardi, L., Minoni, U., and Docchio, F. (1994). A novel, adaptive system for 3-D optical profilometry using a liquid crystal light projector, *IEEE Trans. Instrum. Meas.* **43**, 4, pp. 558–566, doi:10.1109/19.310169.

Self, J., Aiken, C. P., Petibon, R., and Dahn, J. R. (2015). Survey of gas expansion in Li-ion NMC pouch cells, *J. Electrochem. Soc.* **162**, 6, pp. 796–802, doi:10.1149/2.0081506jes.

Winter, B. M., Besenhard, J. O., Spahr, M. E., and Novak, P. (1998). Insertion electrode materials for rechargeable lithium batteries, *Adv. Mater.* **10**, 10, pp. 725–763.

Woodford, W. H., Carter, W. C., and Chiang, Y.-M. (2012). Design criteria for electrochemical shock resistant battery electrodes, *Energy Environ. Sci.* **5**, 7, p. 8014, doi:10.1039/c2ee21874g, http://xlink.rsc.org/?DOI=c2ee21874g.

Yoon, W.-S., Yoon, K., McBreen, J., and Yang, X.-Q. (2006). A comparative study on structural changes of LiCo1/3Ni1/3Mn1/3O2 and LiNi0.8Co0.15Al0.05O2 during first charge using in situ XRD, *Electrochem. Commun.* **8**, pp. 1257–1262, doi:10.1016/j.elecom.2006.06.005.

Zhang, W.-J. (2011). Lithium insertion/extraction mechanism in alloy anodes for lithium-ion batteries, *J. Power Sources* **196**, 3, pp. 877–885, doi:10.1016/j.jpowsour.2010.08.114, http://dx.doi.org/10.1016/j.jpowsour.2010.08.114.

Chapter 7

Every Day a New Battery: Aging Dependence of Internal States in Lithium-ion Cells

Severin Hahn and Kai Peter Birke

7.1 Operation and Degradation Processes in the Electrode State Diagram

7.1.1 *Introduction*

Understanding lithium-ion battery degradation is becoming more important as the service life requirements increase dramatically due to their use in electric vehicles, as compared to consumer electronics. Often, the study of degradation mechanisms aims to reveal methods mitigating the damage. However, as battery aging in general has revealed itself to be inevitable, predicting it's lifetime has gained importance. For example, it is infeasible to test calendar aging when the service life exceeds 10 or even 15 years. Thus, in the automotive industry there is high economic incentive to accurately predict the battery's lifetime due to quality assurance and warranty considerations.

Usually, battery aging is simply expressed as measured; for example, the capacity loss or resistance increase is considered. The underlying degradation mechanisms, however, are diverse. They may fundamentally alter the internal state of the battery and increase, change, passivate or even reverse aging. Hence the aging predictions and mitigation should rely on a model of the internal battery state. This chapter discusses underlying reaction mechanisms and the ways they affect the considered cell state.

First, we will discuss the absolute potentials of the electrodes in a current generation lithium-ion battery cell. With these, the *electrode state diagram* is introduced, in which degradation and other processes can be understood. Then, beginning with simple charge and discharge we develop electrochemical reaction schemes on a theoretical level and evaluate their consequences on the internal cell state. While discussing relevant mechanisms known in literature in this manner, we will encounter different classes of reactions and learn why certain reactions cause cell aging and some do not. This is structured in electrochemical processes changing the internal battery state and degradation of active electrode material. The knowledge of such reactions and their effects is then used to understand the change in absolute electrode potentials, and consequently, the actual origin of capacity fade. We will see that this change can be experimentally detected in aging cells via so-called *differential voltage analysis*. In fact, by investigating these effects we can infer which aging mechanism has degraded the cell without even opening it. Finally, while exploring the aging behavior of a real aged cell with the presented tools, their limitation and deviation from the ideal model are presented. Still, the presented tools provide a strong framework to discuss lithium-ion battery degradation.

This chapter is heavily inspired and based on the work by Ira Bloom (Bloom *et al.*, 2005a,b, 2006, 2010), Jeff R. Dahn (Smith *et al.*, 2011) and Andreas Jossen (Keil and Jossen, 2016; Keil *et al.*, 2016), who developed and introduced this model in multiple papers. Some definitions are directly taken from their publications.

7.1.2 *Absolute potentials and the electrode state diagram*

Electrochemistry is the study of oxidative and reductive processes occurring at the interface between electrodes and electrolytes. Clearly, whether a reaction takes place or not is determined by the energy states of the concerned electrons in the electrode. Some standard electrodes like metals (Zn, Mg, Sn, etc.) have fixed electron energy states that electrochemists have tabulated in the *electrochemical series of metals*. Consequently, by knowing the electron energy difference for our investigated electrode compared to such a fixed electrode, we can classify its environment into being either reductive or oxidative. Handily, electron energy differences are proportional to the measured potential between the two electrodes, which is measured as a *Voltage*. The so-called *absolute potential* of a single electrode is thus always given in reference to a standard electrode. In lithium-ion battery research the chosen reference, by convenience, is lithium metal and consequently absolute potentials are given in *Voltage versus Li/Li^+*. To give some perspective, low and high Voltage versus Li/Li^+ of an electrode describe a reductive and oxidative environment, respectively.

The actual voltage measured between two poles on any battery cell, however, is not absolute. What is measured for such a complete or so-called *full-cell* with both cathode and anode is the difference of the electrode absolute potentials:

$$V_{full\text{-}cell} = V_{cathode} - V_{anode}. \tag{7.1}$$

Here $V_{cathode}$ and V_{anode} denote the absolute electrode potentials of the cathode and anode, respectively. These absolute potentials are also called *half-cell potentials* as they concern a single electrode. The voltage $V_{full\text{-}cell}$ at rest is commonly referred to as *open circuit voltage* or *OCV*.

Current generation intercalation electrodes display varying half-cell potentials depending on the lithiation state as shown in Fig. 7.1(a) and 7.1(b) for a NMC cathode ($Li_y Ni_{1/3} Co_{1/3} Mn_{1/3} O_2$) and a graphite anode ($Li_x C_6$), respectively. Note that the cathode displays a high absolute potential. As we will see later, electrolyte may thus be oxidized on the cathode. On the other hand, anode reactants are reduced. A good visualization [Bloom et al. (2005a)] is achieved when we match the initially uncharged state of a Li-filled

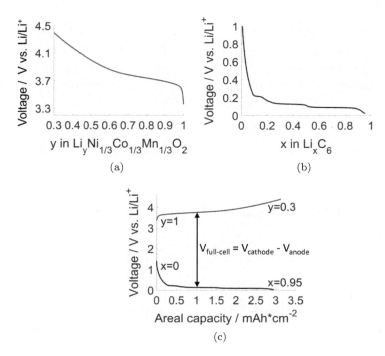

Fig. 7.1 (a) Half-cell potential curve for $Li_y Ni_{1/3} Mn_{1/3} Co_{1/3} O_2$ for different lithiation degree y, (b) half-cell potential curve for $Li_x C_6$ for different lithiation degree x, and (c) electrode state diagram with matched areal capacities.

cathode to a Li-void anode. Most lithium-ion battery cells are assembled in this configuration. Further, we recalculate the lithiation degrees into areal capacities in Ah/m^2 or mAh/cm^2 of the two opposing electrode sheets in the cell. Sometimes, this value is referred to as *electrode loading*. This depends on the film thickness and density of the electrodes in the studied cell and can be calculated and understood by the reader with some assumptions and effort. It should be noted that the areal capacity of a single electrode and its comparative size to the other electrode is an important cell design parameter and is thus variable. In fact, in early cells before other solutions were found the anode was usually chosen to have a larger capacity in order to avoid high lithiation degrees x in graphite. This design helped to reduce lithium plating. Finally, in the resulting representation in Fig. 7.1(c) called an *electrode state diagram*, we can retrace the change in electrochemical environments on the electrodes, for example, during cycling and cell aging. In the following, we will consider different electrochemical reactions and their effect on the state of the battery in this diagram.

7.1.3 Charge and discharge

While charging electrons are pulled from the cathode to the anode through and due to an externally applied power source as sketched in Fig. 7.2(a), the source of electrons is the charge transfer reaction on the cathode surface also releasing Li^+, while the sink of electrons is the corresponding anode reaction consuming Li^+ in the electrolyte.

$$\text{Cathode:} \quad Li(I) \longrightarrow e^- + Li^+$$
$$\text{Anode:} \quad e^- + Li^+ \longrightarrow Li(I)$$

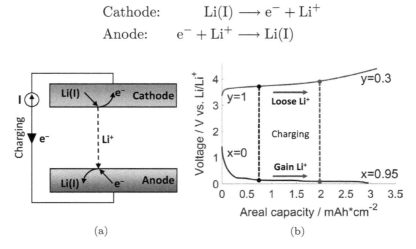

Fig. 7.2 (a) Electrochemical reaction scheme and (b) change to the electrode state diagram for charge.

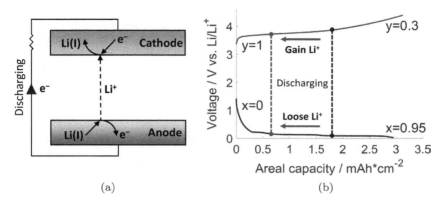

Fig. 7.3 (a) Electrochemical reaction scheme and (b) change to the electrode state diagram for discharge.

Here, Li(I) denotes bound lithium in oxidation state 1 inside an electrode while Li^+ denotes solvated lithium ions. Note that in this and all further depictions, electro-neutrality is assumed and conserved in all parts of the cell. It is straightforward to conclude that the anode gets more lithiated while the cathode is delithiated. Due to the different lithiation degrees, cf. Fig. 7.1, charging moves the internal state simultaneously toward the right on both curves in the electrode state diagram, effectively increasing $V_{cathode}$ and decreasing V_{anode}, as shown in Fig. 7.2(b). Correspondingly, we can deduce that the cell potential $V_{full\text{-}cell}$ increases during charging. Any state of charge of the battery is described by two associated half-cell potentials of the anode and cathode, which is displayed by a connecting dotted line in electrode state diagrams.

Similarly, it is easy to conclude by reversal of flows that discharging has the reverse effect and thus infer the corresponding Fig. 7.3(a) and (b).

7.1.4 *Charge and discharge limits*

Cell charging and discharging are limited by the *maximal* and *minimal cell voltages*, U_{max} and U_{min}, respectively, allowed and chosen by the cell manufacturer. The cell reaches the upper limit while charging when the distance in the curves rises to U_{max} in the electrode state diagram in Fig. 7.4(a). Conversely, the lower limit is reached while discharging when the distance drops to U_{min}. This now conveniently describes the cycling window of the cell. In fact, the area depicted in Fig. 7.4(b) between the cathode and anode voltage limited by U_{min} and U_{max} equates to the *available energy* of

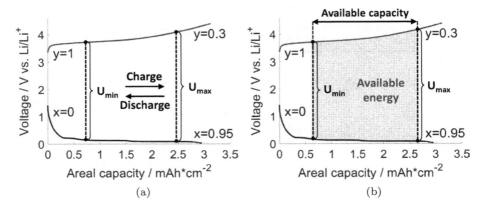

Fig. 7.4 (a) Electrode state diagram with the cycling window boundaries of U_{min} and U_{max} reached by discharging and charging, respectively, and (b) resulting available energy and capacity.

the cell:

$$E_{avail.} = \int_0^{Q_{avail.}} V_{full\text{-}cell} \, dQ$$

$$= \int_0^{Q_{avail.}} (V_{cathode} - V_{anode}) \, dQ. \tag{7.2}$$

Thus, we can later track how degradation mechanisms reduce the energy storage capability in this representation. Similarly, the *available capacity* and its reduction can be read out.

These limits are an important design parameter for a battery cell. Of course, U_{max} and U_{min} should be chosen as high and low, respectively, in order to increase the usable energy of the system. However, there are many degradation phenomena for both sides which have to be considered. Among other effects, at higher U_{max} and thus V versus Li/Li$^+$ on the cathode, the electrolyte is not stable leading to rapid gassing. On the other side, lower U_{min} and correspondingly higher potentials V versus Li/Li$^+$ values on the anode have to be avoided as the commonly used copper current collector may corrode. Choosing the cycling window limits is thus usually a trade-off between the cell energy and its lifetime.

7.1.5 Combined electrode reactions

Figure 7.5 depicts different electrolyte shuttle mechanisms [Smith *et al.* (2011)] where electrolyte species available in the electrolyte, depicted as R, are oxidized in the highly oxidizing environment at high potentials

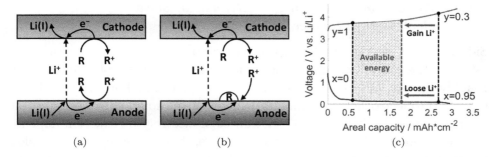

Fig. 7.5 (a) Reaction scheme of reversible self-discharge, (b) reaction scheme of irreversible deposition during self-discharge, and (c) electrode state diagram during self-discharge.

V versus Li/Li$^+$ at the cathode. The additional electron in the cathode will cause the reinsertion of a positive lithium-ion from the electrolyte and therefore retain electro-neutrality inside the active material. Subsequently, the oxidized species can migrate to the anode where it will be reduced. Depending on the shuttled electrolyte species and the specific reaction, the reduced species may be reverted to its initial unoxidized state, as shown in Fig. 7.5(a) or it can deposit on the anode surface as sketched in Fig. 7.5(b) [Smith et al. (2011)]. In both cases, the required electron for reduction stems from delithiation of the anode which releases Li$^+$ into the electrolyte.

Summarizing the reaction mechanisms, the anode is delithiated while the cathode is lithiated by the same amount of charge, causing the half-cell potential to change as depicted in Fig. 7.5(c). Due to the striking similarity in effect on the cell these processes are categorized as self-discharge, cf. Fig. 7.3. Truly, these mechanisms can be thought of as a simple discharge of the cell where the charge (equalizing the lithiation processes) is conducted internally via ionic transfer instead of externally via an electric circuit. Consequently, when stored usable energy is lost the area decreases accordingly in Fig. 7.5(c) and the cell can be simply recharged, which is why self-discharge is generally considered reversible. However, as depicted in Fig. 7.5(b), internal self-discharge may leave irreversible residues on the surface of the anode which may block pores, restrict ionic transport and increase impedance. It is important to note that these reaction products are expected to appear on the macroscopic surface of the anode film close to the separator. Furthermore, electrolyte is consumed, as evident in Fig. 7.5(b) [Smith et al. (2011)]. This can contribute to the drying-out of the electrode, which is a failure mechanism in itself as discussed later.

7.1.6 Anodic side reactions — Growth of solid electrolyte interface (SEI)

Side reactions on the anode involve reduction of electrolyte species. Interestingly, the graphite half-cell potential with prominent plateaus around roughly 90 and 170 mV versus Li/Li$^+$, as depicted in Fig. 7.1 lies clearly outside the electrochemical stability window of the commonly used ethylene carbonate-based solvents. The exact onset of electrolyte reduction depends on the used graphite and electrolyte type, but generally starts below 0.8 to 1.2 V versus Li/Li$^+$ on the anode [Verma et al. (2010)]. Thus, the electrolyte is already reduced during the first lithiation of the anode after assembly, forming a residue on it widely known as a solid electrolyte interface (SEI). This initial SEI growth happening in the first few cycles is a tightly controlled process called formation and is an important and well-studied step in cell manufacturing [German (2015); German et al. (2014)].

As the exact mechanism of SEI growth is heavily disputed in literature, only three prominent proposed processes are presented [Broussely et al. (2001); Ploehn et al. (2004); Li et al. (2015)]. Though not discussed in detail here, it is noted that they lead to a different aging modeling. However, we will see that this does not alter their effect on the electrode state diagram.

The first proposed process [Broussely et al. (2001)] is based on the assumption that the SEI is poorly but finitely electrically conductive. It should be noted that there is stark disagreement to this assumption in literature as inorganic SEI components such as LiF, Li$_2$CO$_3$ and Li$_2$O are known to be insulating [Lin et al. (2016)]. This process, depicted in Fig. 7.6(a), includes the conduction of electrons through the existing SEI, which reduces electrolyte molecules. Then, the anionic electrolyte radicals

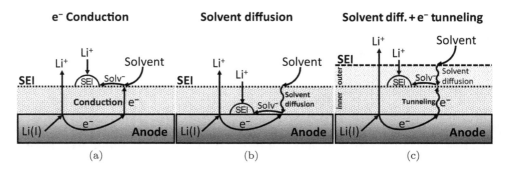

Fig. 7.6 Reaction mechanism schemes for SEI growth driven by (a) electron conduction, (b) solvent diffusion, and (c) combined solvent diffusion and electron tunneling.

react with available Li$^+$ in the electrolyte to form additional SEI. Various chemical reaction pathways for different SEI species and electrolyte species are known to follow this pattern [German (2015)]. Here, we show two electron reduction of ethylene carbonate ($C_3H_4O_3$), producing the principle inorganic SEI compound Li_2CO_3 as an example [German (2015)]:

$$C_3H_4O_3(l) + 2e^- \rightarrow C_3H_4O_3^{2-}$$
$$C_3H_4O_3^{2-} + 2Li^+ \rightarrow Li_2CO_3(s) + C_2H_4 \uparrow$$

Ethylene produced in this reaction is expelled from the reaction site as gas. This may increase internal pressure inside fixed cell formats such as 18650 [Kong et al. (2005)]. In the case of the initial formation cycles in pouch cells, the produced gas is usually removed before the final sealing of the cell. The required electrons for the reaction are again sourced from the delithiation of the anode. The resulting new product species — in our case Li_2CO_3 — precipitates, expanding the existing SEI. As the low electric conduction of the SEI is the limiting factor of further SEI growth the reaction passivates itself and thus mitigates but never completely stops further SEI growth [Broussely et al. (2001)]. Furthermore, the additional SEI growth may restrict ion movement, block pores inside the anode and increase impedance in general. However, in contrast to the residual from irreversible self-discharge, the entire graphite surface, i.e., every single particle in contact with the electrolyte, is covered with a SEI and does not depend on the position in the macroscopic anode film [Verma et al. (2010)]. This is due to the nature of this type of reaction which purely occurs on the anode surface in contact with the electrolyte. In fact, the SEI is built on any lithiated graphite in contact with common electrolytes, regardless of whether there is a corresponding cathode or an external electric connection [Daniel and Besenhard (2011)].

The second proposed process [Ploehn et al. (2004)] suggests a limited diffusion of electrolyte molecules through the SEI, as it is assumed to be porous. This model was motivated by the insulating properties of SEI components mentioned before [Lin et al. (2016)]. The diffused electrolyte molecules are then reduced on the anode surface where a new SEI grows, as Fig. 7.6(b) shows.

The third proposed process [Smith et al. (2011); Li et al. (2015)] also accepts the electronic insulating properties of inorganic SEI components [Lin et al. (2016)] and argues for electron tunneling as the charge transfer mechanism between the graphite surface and an outer porous SEI phase

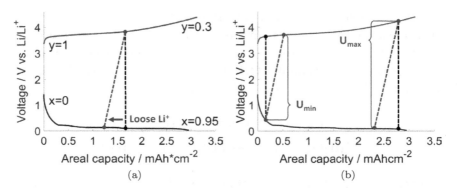

Fig. 7.7 (a) Change in half-cell potential on the anode side due to the loss of Li^+, and (b) corresponding change to the cycling window due to the SEI growth.

where the electrolyte is present [Li et al. (2015)]. As sketched in Fig. 7.6(c) additional SEI is formed within the existing SEI.

The general ramifications of all three processes stay the same. In summary, a passivating SEI film is grown on the anode, gas is produced, electrolyte is consumed, and the anode gets delithiated [Smith et al. (2011)]. This increases the anode potential by moving to the left in Fig. 7.7(a). Most critically, however, the cathode potential is unchanged as the cathode does not take part in this reaction. Thus, a lower anode lithiation state and a consequently higher anode potential are associated to an unchanged cathode lithiation state and potential [Keil and Jossen (2016)]. This inconspicuous change is visualized by a *slanting* of the line associating the anode and cathode voltages in the electrode state diagram [Bloom et al. (2005a)]. Suddenly, when charging to a cell voltage U_{max} by moving to the right as shown in Fig. 7.7(b), we find new cathode and anode potentials associated with this fully charged state. Due to the potential plateau of the highly lithiated graphite anode, the cathode point at the end of charge is hardly changed. The anode end of charge is reached at a much lower lithiation state compared to the initial case. An opposite effect is observed at U_{min}. Due to the steep anode curve the anode end of discharge remains mostly fixed, while the cathode remains less lithiated at U_{min} than in the initial case. These effects may vary with anode and cathode materials with differing corresponding potential curves.

We can immediately observe that both the area and length along the potential curves of the cycling window (and thus the storable energy and capacity, respectively) diminish due to this reaction. In a way, this slanting of the curve is the origin for the observable capacity reduction. This slanting effect and degradation mode is generally referred to as *lithium loss* because a part of the inventory of the so-called *active* or *cyclable lithium* is irreversibly lost into the SEI [Smith et al. (2011)]. As every commercially available cell goes through the process of formation they already display an initial lithium loss and a corresponding slanting in the electrode state diagram. Thus, every commercial battery already is aged when received.

It is important to note that irreversible lithium plating caused by high currents generate the same slanting of the curve because active lithium is lost. Thus, these two degradation mechanisms cause the same effect even though SEI is categorized as calendar aging and Li plating is a cyclic phenomenon in conjunction with a cathode. Additionally, plated Li in contact with both the electrolyte and the anode will immediately react to a new SEI [Petzl et al. (2015)].

SEI may also be created by cyclic aging. Initially, it was argued that diffusion induced strain might crack graphite particles where the SEI might grow on newly exposed graphite surfaces [Christensen and Newman (2006)]. However, it was later clearly shown that graphite was too strong and was predicted (and shown) to fail only at low temperatures and high currents [Takahashi and Srinivasan (2015)]. However, this could not explain the experimentally detected Li loss in cyclic aging experiments at higher temperatures [Ecker et al. (2014)]. Laresgoiti et al. (2015) then made a convincing argument that the SEI itself might crack due to the cyclic volume expansion of the underlying graphite particles. Additional SEI could therefore be grown on the newly exposed surface in the cracks of the SEI.

For further and much more detailed information on SEI formation, structure and composition, the reader is referred to the excellent reviews by Pallavi Verma [Verma et al. (2010)] and Doron Aurbach [Aurbach (2000)].

7.1.7 Cathodic side reactions — Possible formation of solid permeable interface (SPI)

Having studied the previous reaction schemes it might be intriguing to learn that electrochemical oxidation may happen purely on the cathode while maintaining charge neutrality. At least, it has been shown that charged cathodes evolve gas in contact with an electrolyte, without the presence of an anode or an external electric connection [Xiong et al. (2016)]. It should be noted, however, that the following theoretical considerations have not been conclusively verified in experiments. Consider the following reaction scheme in Fig. 7.8. PF_6^- salt anions may be oxidized at the cathode surface to form charge neutral solid PF_6-R oxidation products. Gas may be evolved in

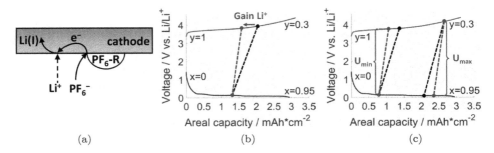

Fig. 7.8 (a) Schematic of purely cathodic oxidation forming SPI on the cathode, (b) correspondingly, the absolute potential on the cathode is reduced due to lithiation. The initial state already displays lithium loss due to formation, and (c) consequent change to the cycling window in the electrode state diagram increasing capacity and storable energy between U_{min} and U_{max}.

this process. Some suggest reaction paths via hydrolysis or LiPF$_6$ instability [Edström et al. (2004)] while others describe oxidation reaction paths as inconclusive [Chagnes and Swiatowsk (2012)]. Possible reaction products like P$_2$O$_5$, Li$_x$PF$_y$, LiF and organic products have been verified in the surface layer of the cathode but they may have initially formed elsewhere in the cell, i.e., not electrochemically on the cathode. This surface layer is often called a *solid permeable interface (SPI)* [Edström et al. (2004)]. If one assumes purely cathodic reactions forming solid oxidation products, as in Fig. 7.8, the electrode sink will be the lithiation of the cathode from the electrolyte [Smith et al. (2011)].

In summary, Li$^+$ and PF$_6$- ions from the electrolyte are consumed while the cathode is covered with a resistance increasing surface film as well as being lithiated. This will cause the cathode potential to decrease as sketched in Fig. 7.8(b), while the anode potential is unaffected [Smith et al. (2011)]. Such reactions can reverse the slanting caused by the SEI formation which is — as discussed before — present in any cell. The less slanted associated half-cell potentials now allow for a larger cycling window, available energy and capacity again. Thus, the aging caused by lithium loss can be reversed by purely cathodic oxidation. Still, as Li$^+$ and PF$_6$-, or rather the LiPF$_6$ salt, is depleted from the electrolyte, its conductivity will decrease and the cell may eventually die [Smith et al. (2011)].

7.1.8 *Transition metal dissolution*

Current generation cathodes suffer from dissolution of their transition metals. Commonly used transition metals like Fe, Co, Mn, Ni and Al have been shown to dissolve in the used organic electrolytes. As the most affected is manganese it is chosen here as an example. At high temperature and cathode absolute potentials in V versus Li/Li$^+$ [Gilbert et al. (2017); Aoshima et al. (2001)] Mn can be dissolved into the electrolyte. While the exact mechanism is disputed [Aoshima et al. (2001); Jang (1996)] it has been suggested to occur as an oxidation state-splitting reaction [Jang (1996)] with subsequent dissolution. One could easily imagine the oxidization state transfer to other transition metals in NMC (Li$_y$Ni$_{1/3}$Mn$_{1/3}$Co$_{1/3}$O$_2$), NCA (Li$_y$Ni$_{0.80}$Co$_{0.15}$Al$_{0.05}$O$_2$) or LFP (LiFePO$_4$) based cathodes.

$$2\text{Mn(III)} \rightarrow \text{Mn(II)} + \text{Mn(IV)}$$

$$\text{Mn(II)} \rightarrow \text{Mn}^{2+}_{\text{solv.}} + 2e^-$$

Here Mn(II), Mn(III), Mn(IV) are manganese in different oxidation states in the cathode and Mn$^{2+}_{\text{solv.}}$ is the dissolved manganese ion. The transition metal dissolution can lead to active material loss and even structural changes in the cathode [Gilbert et al. (2017)]. What active material loss means in terms of the electrode state diagram will be discussed in the following section. As sketched in Fig. 7.9(a), the sink for the produced electrons is again the lithiation of the cathode. The dissolved Mn^{2+} may then migrate to the macroscopic surface of the anode film close to the separator where it is reduced. In fact, transition metals are found in aged lithium-ion cells in the anode [Gilbert et al. (2017)]. It has been suggested that reduced Mn(II) may exist in complexes within the existing SEI catalyzing additional SEI formation [Gilbert et al. (2017)]. The source of electrons for reduction of Mn^{2+} stems from the delithiation of the anode.

To summarize, transition metal dissolution may damage the active material of the cathode, lithiates the cathode and delithiates the anode.

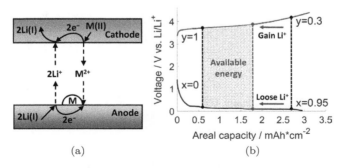

Fig. 7.9 (a) Schematic of transition metal dissolution from the cathode depositing on the anode, and (b) corresponding change in the electrode state diagram discharging the cell.

Figure 7.9 shows that this effect only leads to a classic discharge of the cell, cf. Fig. 7.5, and does not affect the slanting of the curve. This is because transition metal dissolution can be classified as a combined electrode reaction as well with both electrodes taking part.

7.1.9 Loss of active material

All of the reactions discussed above were electrochemical in nature, which changed the battery state. Capacity loss, however, may also be caused by a reduction of storage capability of the electrodes which will be discussed in the following. Active material loss in the context of the electrode state diagram represents the deactivation of storage sites. This can be thought of as a particle detaching itself from the electrode and losing electrical contact. This material is then no longer available to be cycled.

When graphite active material is lost, the anode potential curve is reduced along the capacity axis as shown in Fig. 7.10. To be clear, the curve is only stinted, not cut off. We can see that the unlithiated state remains constant as we assume to lose completely delithiated active material. If this is not the case in reality, we would simply observe the lost Li inside the particle as lithium loss, i.e., as a slanting of the association line. Similarly, the cathode is reduced while fixing the delithiated endpoint, which is also sketched in Fig. 7.10 [Stiaszny et al. (2014a)].

Several effects can cause a loss of active material in the electrodes [Vetter et al. (2005)]. One of these was already encountered in the discussion above

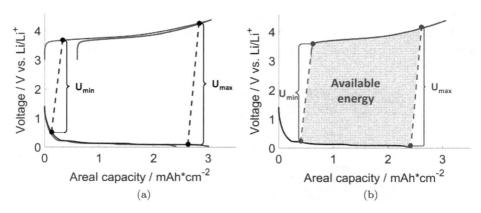

Fig. 7.10 (a) Electrode state diagram showing the initial cycling window after formation and how active material loss affects the half-cell potential curves. The initial state already displays lithium loss due to formation, and (b) consequent change in the remaining cycling window with loss of capacity and energy.

on reaction mechanisms, namely, transition metal dissolution. Upon loss of its transition metals the cathode crystalline structure may be altered to an inactive phase. This inactivation of material equates to an active material loss in the electrode state diagram.

Furthermore, a plethora of effects discussed in literature results in isolation or inactivation of active material [Vetter et al. (2005)]. These include degradation of the additive conductive network, drying out of the electrolyte, corrosion of the current collector (or in general, damage to it), disintegration of the binder, micro and macro cracking of particles, for example, due to diffusion induced strain, exfoliation of graphite, blocking of ionic conduction paths and structural disordering in cathodes.

7.2 Experimental Verification and Analysis Techniques

This section will show how the above concepts can be experimentally verified and used to draw conclusions about the cell aging state. As initially discussed, the measured OCV is the difference between cathode and anode absolute potentials. As seen in Fig. 7.1(b), graphite anodes have characteristic features with steep drops and long plateaus. We can use these characteristics to our advantage. As the information is hard to detect in the full-cell OCV curve, Bloom et al. (2005b) and Dubarry et al. (2011) developed *differential voltage analysis (DVA)* and *incremental capacity analysis (ICA)*, respectively. In these techniques the full-cell OCV curve is differentiated in order to extract the position of electrode characteristics. Note that this can be achieved without opening the actual cell.

DVA: differential voltage: dV/dQ [V/Ah or $mV/(mAh/cm^2)$]

ICA: incremental capacity: dQ/dV [Ah/V or $mAh/(cm^2\ mV)$]

It is simple to conclude that the DVA curve describes the slope of the OCV curve while the ICA curve equals its inverse. Low values in the DVA describe plateaus while peaks in the DVA hint towards drops of the OCV curve. Conversely, low values in the ICA describe drops and troughs describe plateaus. As well resolved OCV measurements are exhaustive and thus hard to come by, usually quasi-OCV measurements are differentiated. These are gathered by a very slow charge or discharge of the investigated cell or electrode, mostly with currents below C/10 [Bloom et al. (2005a, 2005b, 2006, 2010); Keil and Jossen (2016); Keil et al. (2016); Ecker et al. (2014)].

Here, we will focus on the analysis and understanding of the DVA. With the study of relevant literature [Bloom et al. (2005b); Dubarry et al. (2011)],

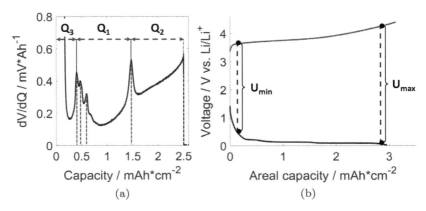

Fig. 7.11 (a) Differential voltage of a NMC/graphite full-cell, where the prominent anode peaks are visible, and (b) corresponding electrode state diagram without assumed Li loss.

an understanding of ICA curves is easily attainable. The two characteristic drops in the anode absolute potential seen in Fig. 7.1(b) are attributed to phase transitions in the anode during which the Li intercalation pattern changes. In the DVA of a full-cell, we can see these phase transitions as peaks as shown in Fig. 7.11. We can now see that the first drop actually contains multiple peaks and thus phase transitions (from phase 1' to IV, to III and to IIL[1]) [Bauer et al. (2016); Cañas et al. (2017)].

When we now compare a DVA curve of a fresh cell to an aged one, we can draw our conclusions about the internal state of the battery. Note that these first examples are produced by changes in the electrode state diagram in the general case, when only the graphite characteristics are available. For this, a NMC/graphite ($Li_yNi_{1/3}Mn_{1/3}Co_{1/3}O_2/Li_xC_6$) system was chosen where cathode features are not pronounced. We will discuss real measurement examples of a battery based on a NCA/graphite chemistry later.

7.2.1 Loss of anode active material

In Fig. 7.12(b), anode active material damage reduces the half-cell potential curve as discussed before. The effect on the measured DVA curve is shown in Fig. 7.12(a). When active material on the anode is lost, the charge stored between phase transition peaks diminishes [Ecker et al. (2014)]. This way, active material loss on the anode can easily be detected and even quantified by the distance Q_1 decreasing. Often, anode material damage may not cause significant loss of capacity due to its very stable plateau at high potential. On the other hand, it may increase the risk of lithium plating

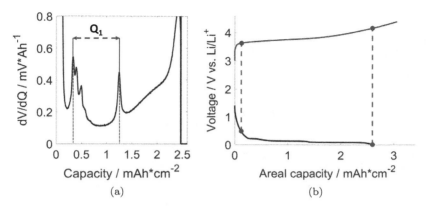

Fig. 7.12 (a) Differential voltage of a NMC/graphite full-cell with simulated loss of anode active material, and (b) corresponding electrode state diagram with an indicated cycling window.

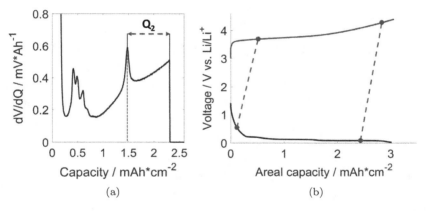

Fig. 7.13 (a) Differential voltage of a NMC/graphite full-cell with simulated loss of active lithium, and (b) corresponding electrode state diagram showing the increased slanting due to lithium loss.

as the remaining anode experiences higher lithiation degrees when charging to U_{max}.

7.2.2 Loss of active lithium

If no anode material damage is present, a loss of active lithium can be detected by a reduction of the distance between the second anode peak and the end of charge, Q_2, in the DVA, as shown in Fig. 7.13(a). The reason for this can be understood by comparing the change of slanting between the electrode state diagrams in Figs. 7.11(b) and 7.13(b).

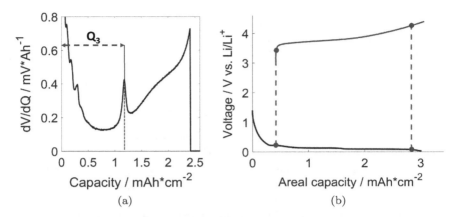

Fig. 7.14 (a) Differential voltage of a NMC/graphite full-cell with simulated loss of cathode active material, and (b) corresponding electrode state diagram with indicated cycling window.

7.2.3 *Loss of cathode active lithium*

When cathode material is lost, the distance of the first and second anode peaks to the end of discharge is reduced in the DVA [Stiaszny *et al.* (2014a)]. The first anode peak might even be shifted out of the active cycling window. Cathode active material damage may be detected by a reduction of the capacity Q_3, as shown in Fig. 7.14(a). The corresponding electrode state diagram in Fig. 7.14(b) shows the remaining cycling window. Also, cathode active material damage can be more easily detected when cathode features are available. Then, the loss of cathode material can be quantified by the distance of cathode peaks to the end-of-charge, which will be shown later.

7.2.4 *The principle of limitation*

When multiple aging mechanisms are present at the same time they might not linearly add up. Consequently, detection and quantization by DVA become complex. Figure 7.15 shows the grave lithium loss mentioned in Sec. 7.2.2, cf. Fig. 7.13, with superimposed cathode material damage. Interestingly, the cell is hardly affected by the additional active material damage. This is because the loss of active material would affect a region on the cathode which is not reached within the cycling window caused by the lithium loss. When multiple effects are present, a simple diagnosis via a non-intrusive DVA might not be enough to quantify aging mechanisms nor the limiting factor. In this case, however, the cell would be limited by active lithium loss which can be understood in Fig. 7.15(b).

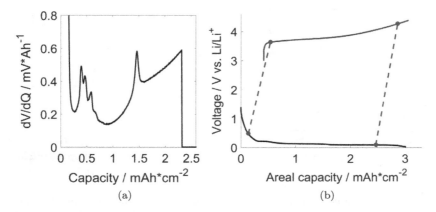

Fig. 7.15 (a) Differential voltage of a NMC/graphite full-cell with Li loss from Fig. 7.13 with superimposed cathode active material damage, and (b) corresponding electrode state diagram with indicated cycling window.

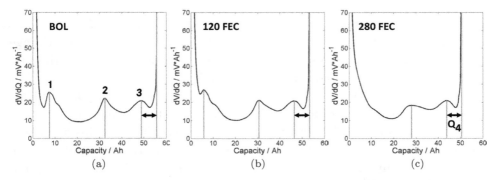

Fig. 7.16 Differential voltage analysis of a NCA/graphite battery system at (a) beginning of life, (b) after 140 FEC, and (c) after 280 FEC of combined cyclic aging at 10°C and between −15 and 45°C. DVA curves are recorded during repeatable parameter tests with a nominal C/24 discharge current.

7.2.5 Example of an aged cell

Figure 7.16(a) shows an unaged cell with a $Li_yNi_{0.80}Co_{0.15}Al_{0.05}O_2$ cathode and graphite anode. The first two peaks are associated to the graphite anode while the last peak belongs to the NCA cathode. After cyclic aging for 120 *Full Equivalent Cycles (FEC)*, we observe the DVA curve in Fig. 7.16(b) followed by a DVA after 280 FEC in Fig. 7.16(c).

Unfortunately, we lose sight of the first peak of the graphite anode during aging, which we could use to analyze active material loss of the anode as in Section 7.2.1. Clearly, however, we lose sight of this peak due to grave active

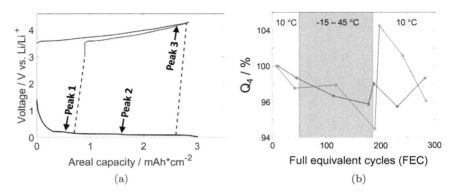

Fig. 7.17 (a) Assumed schematic of an electrode state diagram for the aging of a NCA/graphite cell with indicated characteristic peak positions 1 and 2 for the graphite anode and peak position 3 for the NCA cathode. NCA half-cell data was sourced from literature [Keil and Jossen (2016)] and (b) quantitative evaluation over aging of the DVA feature Q4 in Fig. 7.16 during combined cyclic aging at 10°C and between −15 and 45°C as indicated. Data for two separate battery systems is displayed.

material damage to the cathode which limits the cell. Figure 7.17(a) shows an assumed sketch of the electrode state diagram and its assumed change during aging. This illustrates how we cannot reach the position of the first anode peak as it lies outside the cycling window.

As hinted earlier, cathode material damage can be quantified by analyzing the capacity difference from characteristic cathode features, in this case Peak 3 in Fig. 7.17(a), to the end of charge. This has been denoted Q_4 in Fig. 7.16. The quantitative analysis of the relative to initial value of Q_4 over its lifetime in combined cyclic aging at 10°C and between −15 and 45°C is shown in Fig. 7.17(b) for two battery systems. This example shows that quantitative DVA experiments are fragile, for example to resistance, and data might not be highly reliable. However, if one assumes an average reduction of Q_4 to 97% in Fig. 7.17(b), a cathode material loss of 24% for the entire cathode may be extrapolated. This evaluation also supports the qualitative arguments for the cathode limitation listed above.

7.2.6 Inhomogeneities and limitations in real cells

The reader might have recognized that in real aged cells, i.e., in Fig. 7.16, DVA-peaks change shape or reduce in height while they remain stable in generically produced DVA curves, i.e., Fig. 7.13. It was suggested early on [Bloom et al. (2006)] that part of this could be due to the co-intercalation of electrolyte. More pronounced reduction in peak height, similar to Fig. 7.16, has sometimes been falsely identified as active material damage. In reality, however, this has been shown [Sieg et al. (2016)] to be caused by inhomogeneous aging in the plane of the electrode. The model introduced in this chapter may only ideally describe small sections of opposing cathode and anode electrodes. In fact, lithium loss can be more severe in certain parts of a large anode [Sieg et al. (2016)]. The result would be an inhomogeneous lithiation and aging state in the electrode plane. Actually, graphite anodes can coexist for months in different lithiation states in the electrode plane without significant intercharging due to the vast plateaus of similar potential [Lewerenz et al. (2017)], cf. Fig. 7.1(b). A certain slanting might thus only be representative for a certain part of an anode, while another part displays a different lithium loss. Then, when distinct phase transitions peaks in the DVA are separately encountered at slightly offset capacities starting from a boundary U_{min} or U_{max}, the peak appearance widens and its size decreases. For example, while discharging from U_{max} the second graphite peak is encountered earlier in Fig. 7.13 than in Fig. 7.11. Similarly, other degradation mechanisms other than lithium loss may also be localized in certain parts of the electrode plane.

This presents both an opportunity and a challenge for the application of DVA for aging analysis in unopened cells. In a way, electrode aging inhomogeneity may be qualified by this analysis for unopened full cells: when peak heights are reduced, some aging mechanism is inhomogeneously distributed in the cell, which suggests inhomogeneous loads or temperatures in the aging process. For example, battery cooling systems aim to achieve homogeneous temperature distributions in a cell which can be qualified in this way. Still, attribution of the inhomogeneity to a specific aging mechanism or its quantization are difficult by DVA. Furthermore, quantitative analysis of aging mechanisms by DVA becomes highly inaccurate with inhomogeneity present. This is a fundamental limitation of DVA in large cells.

7.3 Conclusion

This chapter described different kind of reactions that affect the internal state of the battery. Combined electrode reactions, such as self-discharge and transition metal dissolution cause a simple discharge of the cell, without affecting the cycling window. Anodic side reactions lead to a loss of active lithium slanting the association curve in the electrode state diagram. On the other hand, cathodic side reactions may theoretically reverse capacity aging due to slanting by sourcing for new active lithium from the electrolyte. Deactivation of storage sites is observed as a reduction of half-cell curves along the capacity axis in the electrode state diagram. These aging effects can be detected, differentiated and even sometimes quantified by DVA.

Often in battery development, the external measured cell voltage at the taps of a battery is thought of and used as a state defining parameter. The derived concept of a State of Charge (SoC) even describes itself as a state. The SoC is usually defined to be 0% at U_{min}, 100% at U_{max} and changing inbetween by differing capacity. As we saw, however, U_{min} and U_{max} positions in the electrode state diagram — and thus the entire definition of SoC — may be altered. Based on the discussion in this chapter, it is very important to realize that these do not describe a fixed state over the lifetime of a cell. In fact, these cell states change, albeit slightly, every day.

Moreover, many important properties depend on the state of lithiation of the electrodes which thereby change due to aging:

$\Rightarrow$ Electrode diffusion coefficients [Persson et al. (2010)]

                                              affecting resistance.

$\Rightarrow$ Open circuit voltage (OCV)

                                        affecting battery management.

$\Rightarrow$ Combustibility [Yamaki et al. (2003)]

                                                        affecting safety.

$\Rightarrow$ Electrode absolute potentials

                                           affecting further aging.

$\Rightarrow$ Active material volume expansion [Laresgoiti et al. (2015)]

                                           affecting further aging.

The effect of a changing electrode state diagram should be considered when studying these topics, which is already but rarely being done. For example, calendar aging of lithium-ion batteries has been shown to depend strongly on the graphite anode absolute potential [Keil and Jossen (2016); Keil et al. (2016); Ecker et al. (2014)]. This gives a complex implicit relationship between aging and the consequent electrode state diagram change, thus affecting further degradation [Keil et al. (2016)]. In another study, however, the change of the OCV curve over its lifetime was empirically measured without going into detail on the origins in the electrode state diagram, which might have been beneficial [Farmann and Sauer (2017)]. With a clearer understanding of these effects, aging predictions, and possibly the lithium-ion cell's lifetime itself, may be improved in the future.

In summary, it is very useful to think of a lithium-ion battery cell operation in terms of absolute potentials and the electrode state diagram. Of course, the reaction schemes above are far more complex and diverse than what was presented here because this chapter focuses solely on their consequences on the internal electrode state. For example, SEI species and thus reaction pathways are highly diverse. However, even in this simplified model, many material and cell design choices may be explained. Furthermore, any degradation in capacity and consequently energy can be understood and even quantitatively measured using these powerful tools. The DVA in particular offers great insights into the cell aging state without opening the cell. In the future, this may be detected by smart battery management systems that can alter the operation strategy to prevent further damage [Sieg et al. (2016)]. In most aging experiments, the DVA is worth its significant measurement time investment. Where the DVA falls short due to effects discussed above, more extensive post-mortem studies can still recover the electrode state diagram information by investigating half-cells directly [Stiaszny et al. (2014a,b)].

Bibliography

Aoshima, T., Okahara, K., Kiyohara, C., and Shizuka, K. (2001). Mechanisms of manganese spinels dissolution and capacity fade at high temperature, *J. Power Sources* **97–98**, pp. 377–380, doi:10.1016/S0378-7753(01)00551-1.

Aurbach, D. (2000). Review of selected electrode–solution interactions which determine the performance of Li and Li-ion batteries, *J. Power Sources* **89**, 2, pp. 206–218, doi:10.1016/S0378-7753(00)00431-6.

Bauer, M., Wachtler, M., Stöwe, H., Persson, J. V., and Danzer, M. A. (2016). Understanding the dilation and dilation relaxation behavior of graphite-based lithium-ion cells, *J. Power Sources* **317**, pp. 93–102, doi:10.1016/j.jpowsour.2016.03.078.

Bloom, I., Christophersen, J., and Gering, K. (2005a). Differential voltage analyses of high-power lithium-ion cells, *J. Power Sources* **139**, 1–2, pp. 304–313, doi:10.1016/j.jpowsour.2004.07.022.

Bloom, I., Christophersen, J. P., Abraham, D. P., and Gering, K. L. (2006). Differential voltage analyses of high-power lithium-ion cells, *J. Power Sources* **157**, 1, pp. 537–542, doi:10.1016/j.jpowsour.2005.07.054.

Bloom, I., Jansen, A. N., Abraham, D. P., Knuth, J., Jones, S. A., Battaglia, V. S., and Henriksen, G. L. (2005b). Differential voltage analyses of high-power, lithium-ion cells, *J. Power Sources* **139**, 1–2, pp. 295–303, doi:10.1016/j.jpowsour.2004.07.021.

Bloom, I., Walker, L. K., Basco, J. K., Abraham, D. P., Christophersen, J. P., and Ho, C. D. (2010). Differential voltage analyses of high-power lithium-ion cells, *J. Power Sources* **195**, 3, pp. 877–882, doi:10.1016/j.jpowsour.2009.08.019.

Broussely, M., Herreyre, S., Biensan, P., Kasztejna, P., Nechev, K., and Staniewicz, R. (2001). Aging mechanism in Li-ion cells and calendar life predictions, *J. Power Sources* **97-98**, pp. 13–21, doi:10.1016/S0378-7753(01)00722-4.

Cañas, N. A., Einsiedel, P., Freitag, O. T., Heim, C., Steinhauer, M., Park, D.-W., and Friedrich, K. A. (2017). Operando X-ray diffraction during battery cycling at elevated temperatures: A quantitative analysis of lithium-graphite intercalation compounds, *Carbon* **116**, pp. 255–263, doi:10.1016/j.carbon.2017.02.002.

Chagnes, A. and Swiatowsk, J. (2012). Electrolyte and solid-electrolyte interphase layer in lithium-ion batteries, in I. Belharouak (ed.), *Lithium-ion Batteries — New Developments* (InTech), ISBN 978-953-51-0077-5, doi:10.5772/31112.

Christensen, J. and Newman, J. (2006). Stress generation and fracture in lithium insertion materials, *J. Solid State Electrochem.* **10**, doi:10.1007/s10008-006-0095-1.

Daniel, C. and Besenhard, J. O. (2011). *Handbook of Battery Materials* (Wiley-VCH Verlag GmbH & Co. KGaA, Weinheim, Germany), ISBN 9783527637188, doi: 10.1002/9783527637188.

Dubarry, M., Truchot, C., Liaw, B. Y., Gering, K., Sazhin, S., Jamison, D., and Michelbacher, C. (2011). Evaluation of commercial lithium-ion cells based on composite positive electrode for plug-in hybrid electric vehicle applications. Part II. Degradation mechanism under 2C cycle aging, *J. Power Sources* **196**, 23, pp. 10336–10343, doi:10.1016/j.jpowsour.2011.08.078.

Ecker, M., Nieto, N., Käbitz, S., Schmalstieg, J., Blanke, H., Warnecke, A., and Sauer, D. U. (2014). Calendar and cycle life study of Li(NiMnCo)O_2-based 18650 lithium-ion batteries, *J. Power Sources* **248**, pp. 839–851, doi:10.1016/j.jpowsour.2013.09.143.

Edström, K., Gustafsson, T., and Thomas, J. O. (2004). The cathode–electrolyte interface in the Li-ion battery, *Electrochim. Acta* **50**, 2–3, pp. 397–403, doi:10.1016/j.electacta.2004.03.049.

Farmann, A. and Sauer, D. U. (2017). A study on the dependency of the open-circuit voltage on temperature and actual aging state of lithium-ion batteries, *J. Power Sources* **347**, pp. 1–13, doi:10.1016/j.jpowsour.2017.01.098.

German, F., Hintennach, A., LaCroix, A., Thiemig, D., Oswald, S., Scheiba, F., Hoffmann, M. J., and Ehrenberg, H. (2014). Influence of temperature and upper cut-off voltage on the formation of lithium-ion cells, *J. Power Sources* **264**, pp. 100–107, doi:10.1016/j.jpowsour.2014.04.071.

German, F. (2015). Untersuchungen zur SEI-bildung und optimierung der formation an lithium-ionen voll- und halbzellen, Ph.D. thesis, Karlsruhe Institute of Technology.

Gilbert, J. A., Shkrob, I. A., and Abraham, D. P. (2017). Transition metal dissolution, ion migration, electrocatalytic reduction and capacity loss in lithium-ion full cells, *J. Electrochem. Soc.* **164**, 2, pp. A389–A399, doi:10.1149/2.1111702jes.

Jang, D. H. (1996). Dissolution of spinel oxides and capacity losses in 4 V Li/$Li_xMn_2O_4$ cells, *J. Electrochem. Soc.* **143**, 7, p. 2204, doi:10.1149/1.1836981.

Keil, P. and Jossen, A. (2016). Calendar aging of NCA lithium-ion batteries investigated by differential voltage analysis and coulomb tracking, *J. Electrochem. Soc.* **164**, 1, pp. A6066–A6074, doi:10.1149/2.0091701jes.

Keil, P., Schuster, S. F., Wilhelm, J., Travi, J., Hauser, A., Karl, R. C., and Jossen, A. (2016). Calendar aging of lithium-ion batteries, *J. Electrochem. Soc.* **163**, 9, pp. A1872–A1880, doi:10.1149/2.0411609jes.

Kong, W., Li, H., Huang, X., and Chen, L. (2005). Gas evolution behaviors for several cathode materials in lithium-ion batteries, *J. Power Sources* **142**, 1–2, pp. 285–291, doi:10.1016/j.jpowsour.2004.10.008.

Laresgoiti, I., Käbitz, S., Ecker, M., and Sauer, D. U. (2015). Modeling mechanical degradation in lithium-ion batteries during cycling: Solid electrolyte interphase fracture, *J. Power Sources* **300**, pp. 112–122, doi:10.1016/j.jpowsour.2015.09.033.

Lewerenz, M., Münnix, J., Schmalstieg, J., Käbitz, S., Knips, M., and Sauer, D. U. (2017). Systematic aging of commercial LiFePO$_4$ — Graphite cylindrical cells including a theory explaining rise of capacity during aging, *J. Power Sources* **345**, pp. 254–263, doi:10.1016/j.jpowsour.2017.01.133.

Li, D., Danilov, D., Zhang, Z., Chen, H., Yang, Y., and Notten, P. H. L. (2015). Modeling the SEI-formation on graphite electrodes in LiFePO$_4$ batteries, *J. Electrochem. Soc.* **162**, 6, pp. A858–A869, doi:10.1149/2.0161506jes.

Lin, Y.-X., Liu, Z., Leung, K., Chen, L.-Q., Lu, P., and Qi, Y. (2016). Connecting the irreversible capacity loss in Li-ion batteries with the electronic insulating properties of solid electrolyte interphase (SEI) components, *J. Power Sources* **309**, pp. 221–230, doi:10.1016/j.jpowsour.2016.01.078.

Persson, K., Sethuraman, V. A., Hardwick, L. J., Hinuma, Y., Meng, Y. S., van der Ven, A., Srinivasan, V., Kostecki, R., and Ceder, G. (2010). Lithium diffusion in graphitic carbon, *J. Phys. Chem. Lett.* **1**, 8, pp. 1176–1180, doi:10.1021/jz100188d.

Petzl, M., Kasper, M., and Danzer, M. A. (2015). Lithium plating in a commercial lithium-ion battery — A low-temperature aging study, *J. Power Sources* **275**, pp. 799–807, doi:10.1016/j.jpowsour.2014.11.065.

Ploehn, H. J., Ramadass, P., and White, R. E. (2004). Solvent diffusion model for aging of lithium-ion battery cells, *J. Electrochem. Soc.* **151**, 3, p. A456, doi:10.1149/1.1644601.

Sieg, J., Behrends, P., Mitsch, T., Nuhic, A., Lay, J., Spier, B., and Sauer, D. (2016). Local degradation and differential voltage analysis of inhomogeneously aged lithium-ion pouch cell, Poster presented at 15th Ulm ElectroChemical Talks UECT, June, Blaubaeuren, Germany, pp. 20–21.

Smith, A. J., Burns, J. C., Xiong, D., and Dahn, J. R. (2011). Interpreting high precision coulometry results on Li-ion cells, *J. Electrochem. Soc.* **158**, 10, p. A1136, doi:10.1149/1.3625232.

Stiaszny, B., Ziegler, J. C., Krauß, E. E., Schmidt, J. P., and Ivers-Tiffée, E. (2014a). Electrochemical characterization and post-mortem analysis of aged LiMn$_2$O$_4$-Li(Ni$_{0.5}$Mn$_{0.3}$Co$_{0.2}$)O$_2$/graphite lithium-ion batteries. Part I: Cycle aging, *J. Power Sources* **251**, pp. 439–450, doi:10.1016/j.jpowsour.2013.11.080.

Stiaszny, B., Ziegler, J. C., Krauß, E. E., Zhang, M., Schmidt, J. P., and Ivers-Tiffée, E. (2014b). Electrochemical characterization and post-mortem analysis of aged LiMn$_2$O$_4$–NMC/graphite lithium-ion batteries. Part II: Calendar aging, *J. Power Sources* **258**, pp. 61–75, doi:10.1016/j.jpowsour.2014.02.019.

Takahashi, K. and Srinivasan, V. (2015). Examination of graphite particle cracking as a failure mode in lithium-ion batteries: A model-experimental study, *J. Electrochem. Soc.* **162**, 4, pp. A635–A645, doi:10.1149/2.0281504jes.

Verma, P., Maire, P., and Novák, P. (2010). A review of the features and analyses of the solid electrolyte interphase in Li-ion batteries, *Electrochim. Acta* **55**, 22, pp. 6332–6341, doi:10.1016/j.electacta.2010.05.072.

Vetter, J., Novák, P., Wagner, M. R., Veit, C., Möller, K.-C., Besenhard, J. O., Winter, M., Wohlfahrt-Mehrens, M., Vogler, C., and Hammouche, A. (2005). Ageing mechanisms in lithium-ion batteries, *J. Power Sources* **147**, 1–2, pp. 269–281, doi:10.1016/j.jpowsour.2005.01.006.

Xiong, D. J., Ellis, L. D., Petibon, R., Hynes, T., Liu, Q. Q., and Dahn, J. R. (2016). Studies of gas generation, gas consumption and impedance growth in Li-ion cells with carbonate or fluorinated electrolytes using the pouch bag method, *J. Electrochem. Soc.* **164**, 2, pp. A340–A347, doi:10.1149/2.1091702jes.

Yamaki, J.-I., Baba, Y., Katayama, N., Takatsuji, H., Egashira, M., and Okada, S. (2003). Thermal stability of electrolytes with Li_xCoO_2 cathode or lithiated carbon anode, *J. Power Sources* **119–121**, pp. 789–793, doi:10.1016/S0378-7753(03)00254-4.

Chapter 8

Thermal Propagation

Sascha Koch

8.1 Introduction

Manufacturing errors, misuse, abuse or even poorly chosen operating conditions can result in cell internal chemical reactions between different components and materials. These reactions usually lead to heat and gas generation and can show a runaway behavior which climaxes in a catastrophic cell reaction and propagation. This chapter describes the processes, possible tests and influencing variables of this behavior, specifically focusing on lithium-ion cells and batteries.

8.2 Process of Thermal Propagation

Often described as a thermal event, the thermal propagation process within a lithium-ion battery is actually the chain reaction of single cell thermal runaway events triggering each other. To understand this propagation process it is crucial to first understand the mechanisms and reactions taking place during the thermal runaway within a single cell.

The following sections will describe the thermal runaway in a single cell and then move on to the thermal runaway propagation, the so-called thermal propagation.

8.2.1 *Thermal runaway*

The mainly exothermic chemical reactions between the electrodes and the electrolyte of a lithium-ion cell are called a thermal runaway. Each of these reactions introduce more heat into the thermal body of the cell, pushing its temperature above the trigger level of the next reaction. With the amount of heat added to the system being higher than the dissipated heat,

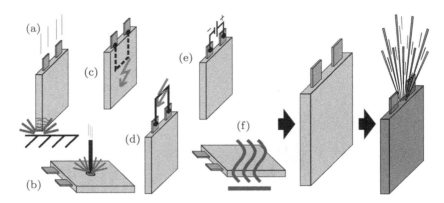

Fig. 8.1 Trigger mechanisms able to induce a thermal runaway in a lithium-ion cell. Depending on the cell type not all trigger mechanisms inevitably have to lead to a thermal runaway.

a self-catalytic behavior can be observed justifying the denotation thermal runaway.

Figure 8.1 illustrates some common causes triggering a thermal runaway [Tobishima and Yamaki (1999)]:

(a) Crash/crush
(b) Penetration
(c) Internal short circuit
(d) External short circuit
(e) Overcharge
(f) Overheat

Some of these causes lead to others. A crush/crash and a penetration usually results in a shift in the cell stack or punctuation of the separator, which leads to an electrical connection of the anode and cathode layers and therefore producing an internal short circuit. The electrical energy gets transferred into heat during the discharge and can cause the cell to overheat. The external short circuit produces high currents which again can cause excess heat within the cell, due to the internal resistance leading the cell to overheat. During overcharging a depletion of lithium in the cathode occurs leading up to several exothermic reactions such as electrolyte oxidation [MacNeil and Dahn (2002); Kumai et al. (1999)] and decomposition of the cathode material [Dahn et al. (1994); Biensa et al. (1999)]. Thus, all of the above primary trigger mechanisms lead to an increase in cell temperature, which itself can cause a meltdown of the separator at temperatures of $T_{\text{meltdown}} \approx 125°C \leq 155°C$, if not already penetrated.

The single processes and reactions taking place during a thermal runaway have been studied on a Li(Ni$_x$Co$_y$Mn$_z$)O$_2$ (NCM) lithium-ion cell with an accelerating rate calorimeter (ARC) [Feng et al. (2014)]. It starts with the decomposition of the solid electrolyte interface (SEI) layer and the self discharge of the cathode from $T \approx 60°C$ on, and is followed by reactions of the anode and electrolyte starting from $T \approx 70°C$. The separator melts between $T \approx 120°C$ and $T \approx 160°C$, resulting in micro short circuits and rising self-heating rates of the cell. Finally, after a hard short circuit happens at $T \approx 230°C$, resulting from the breakup of the separator, and shortly after at $T \approx 250°C$, major exothermic reactions occur due to the cathode, electrolyte and binder decomposition, which lead to the highest self-heating rates during the thermal runaway [Feng et al. (2014); Kim et al. (2007); Spotnitz and Franklin (2003)].

The decomposition of the electrolyte and its reactions with the electrode materials results in gas generation. This gas mainly consists of carbon monoxide, carbon dioxide, hydrogen and other carbon hydrogens with a flammability at fuel concentrations in the air from $c_{\text{flam}} \approx 4\%$ to $c_{\text{flam}} \approx 40\%$ [Mikolajczak et al. (2011); Ponchaut et al. (2015)].

The released energy during a thermal runaway reaction of a lithium-ion cell exceeds its electrochemically stored energy by a factor of $7 \leq F \leq 10$. The electrochemically saved energy gets transferred into heat during a thermal runaway and therefore

$$E_{electrical} \approx E_{thermal}, \tag{8.1}$$

the rest of the energy stays chemically bound in the released venting gas, as long as it is not ignited [Doughty (2012)].

As mentioned above, the single reactions taking place at different temperatures during the thermal runaway are mostly exothermic, each producing enough heat to start the next reaction with a higher trigger temperature. The only way to stop this chain reaction is to direct the generated heat out of the system and not let the cell reach the trigger temperatures for new thermal runaway reactions [Tobishima and Yamaki (1999)].

8.2.2 *Propagation*

If cells are stacked together in modules with good thermal coupling, a single cell thermal runaway described in Sec. 8.2.1 can produce enough excess heat to trigger a thermal runaway in adjacent cells. The spread of this effect is called thermal propagation.

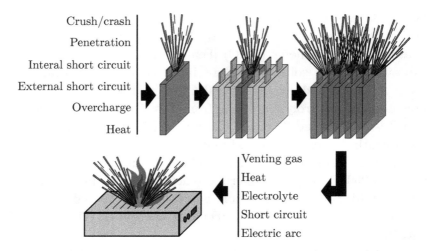

Fig. 8.2 Process of thermal propagation. After triggering the thermal runaway in a single cell it produces enough heat to trigger adjacent cells into a thermal runaway. This propagation can lead to the destruction of the housing integrity.

Figure 8.2 shows the process of thermal propagation within a lithium-ion battery. Starting with one of the initial thermal runaway triggers illustrated in Fig. 8.1, a single cell undergoes a thermal runaway. The chemical reactions taking place produce excess heat which leads to an increase in cell temperature. Because of the thermal coupling of the cells, either by directly touching each other or through a common heat sink or cooler, the heat of the damaged cell is transferred onto adjacent cells. As heat can trigger a thermal runaway, the adjacent cells undergo their own thermal runaway reaction if the necessary temperatures are reached. Driven by the spread of heat throughout the battery and the ongoing heat input through cells undergoing thermal runaway, the propagation can reach all thermally coupled cells. Based on the number of cells, arrangement in modules and the thermal coupling between the cells and modules, this propagation can take hours. Experiments showed a propagation time from cell to cell of about $t_{\text{prop}} = 10\,\text{s} \leq 60\,\text{s}$ for pouch cells at the beginning of the thermal propagation. The longer the propagation goes on and the more preheated adjacent cells become, the faster the ignition of these cells occur.

At the beginning of the thermal propagation process the battery housing has to cope with venting gas. For the gas generation a volume of $V_{\text{gas}} \approx 2 \leq 3\,\text{l}$ of produced gas per Ah capacity can be expected. The gas of a single cell is usually generated and released in a time frame of $t \approx 10 \leq 30\,\text{s}$. In addition to the gas a huge amount of thermal energy is released.

As mentioned in Sec. 8.2.1 the amount of thermal energy set free by a single cell is roughly equivalent to its capacity. Measurements during previous experiments showed temperatures of $T \approx 600 \leq 950°C$ inbetween two adjacent 60 Ah pouch cells. It is possible that the electrolyte gets ejected together with the gas or leaks from broken cell encapsulations during the thermal propagation.

8.2.3 *Resulting effects*

The course of events during thermal propagation described in Sec. 8.2.2 leads to several impacts on the battery housing and possibly its environment. Some of these impacts are [Wang *et al.* (2012)]:

- Increase of pressure
- Flying parts
- Cell voltage drop
- Electric arcs and sparks
- Fire
- Jet flames and explosion

To prevent further hazardous situations and damage, a battery housing or system should be able to cope with these effects. To avoid uncontrolled rupture of the battery housing an overpressure valve or vent opening should be installed. However, this also leads to an opening in the housing and therefore the possibility of ejecting sparks together with the venting gas. As described in Sec. 8.2.1 the venting gas is highly flammable and the ejected glowing parts have the potential to ignite the gas once mixed with air. Furthermore, a drop or even loss of cell voltage can cause trouble for any functional safety mechanisms powered by the cells. Once in a thermal runaway, a cell can no longer deliver electrical energy.

The last three bullet points in the list are possibly the most severe and dangerous effects. During a thermal runaway when several cells are damaged but many are still intact, the possibility of electrical arcs and sparks is at its highest. Due to the released heat and gas, electrically insulating parts often get partially or completely destroyed. Electrical potentials of up to several hundred volts from yet healthy cells, the missing insulation and hot gas can facilitate the ignition of an electric arc. Once the arc is formed only a low voltage is necessary to maintain it. The arc plasma can reach temperatures of up to several thousand degrees [Lee (1982)] and is therefore hot enough to melt most materials used within a battery. This can lead to uncontrolled and dangerous openings of a battery housing.

If the venting gas outside the battery gets ignited, either by flying glowing parts, electric sparks or arcs, the energy chemically bound in the gas is released and can cause damage to the environment of the battery. Depending on how the gas is released from the battery housing, jet flames can form or explosions can occur if a larger amount of gas gets trapped in a closed room together with the right amount of oxygen.

8.3 Testing

In Sec. 8.2 possible trigger mechanisms for a thermal runaway and propagation as well as their impacts have been described. Most of them strongly depend on the cell type, size and the battery system design. Therefore it is crucial to test newly developed lithium-ion batteries and learn how the system reacts and behaves if a thermal runaway occurs.

8.3.1 *Relevance*

To guide manufacturers in how and what to test, several standards and norms have been developed. While there are test descriptions in norms and standards for different electrical, mechanical and thermal misuse and abuse tests on cell [IEC (2016)] and battery levels [Underwriters Laboratories Inc. (2012)], not many cover thermal propagation. A recently published test description is included in the Sandia Report 2017 [Orendorf et al. (2017)] mentioned under "3.4. Failure Propagation Test" based on a paper about failure propagation [Lamb et al. (2015)] and an older Sandia Report [Orendorff et al. (2014)]. Another test dealing with thermal propagation on a battery or module level is published in SAE J 2464 under Sec. 4.4.5's "Passive Propagation Resistance Test" [SAE-International (2009)]. None of these tests describe a pass or fail criteria; they are solely designed to evaluate the system behavior.

8.3.2 *Trigger methods*

To be able to adequately investigate the behavior of a lithium-ion battery system undergoing thermal propagation, a reliable and reproducible trigger mechanism for a thermal runaway event is needed. In the "Passive Propagation Resistance Test" [SAE-International (2009)] introduced in Sec. 8.3.1, overheating is mentioned as a trigger method, but the option to use other methods is also given. In the "Failure Propagation Test" [Orendorf et al. (2017)] it is required to first find a suitable trigger for the device under test (DUT). Thus, it is important that the trigger actually leads to a thermal runaway within a single cell and is applicable to the DUT without altering its thermal or physical behavior too much. Experience from previous tests and experiments showed the following trigger methods as feasible:

- Overcharging
- Nail penetration
- Overheating

How each of these trigger methods act on the cell is described in Sec. 8.2.1. Overcharging as an abuse test and therefore trigger method can be found in SAE J 2464 Sec. 4.5.2's "Overcharge Test" [SAE-International (2009)]. To speed up the test, charging should be done at the highest rated current of the cell. Depending on the cell type, overcharging up to a State of Charge (SoC) of $\approx 150\% \leq 200\%$ can be necessary to initiate a thermal runaway. If no thermal runaway occurs at SoC = 200% this method can be considered as unfit. An advantage of using overcharging as a trigger method is how easily accessible the cell tabs usually are, allowing for only little changes to the thermal properties of the battery setup. A big disadvantage is the additional electrical energy that gets introduced to the system by overcharging. This additional energy represents an abnormal system state and could lead to a more severe reaction than to be expected during normal operation cell failure. To limit this effect, the charging source should be switched off as soon as thermal runaway reactions are detected.

Nail penetration is described in SAE J 2464 Sec. 4.3.3's "Penetration Test" [SAE-International (2009)]. A conductive steel rod with a diameter of $d = 3\,\mathrm{mm}$ with a sharp point (e.g., a nail) is driven into the cell. If no runaway reaction is observed after penetrating the cell fully, this trigger method can be seen as unfit. Advantages of this method is that there is no change in the systems' energy and no change to the thermal properties. A clear disadvantage is the necessary drive mechanism. This is usually a hydraulic or pneumatic cylinder which is rather big and needs mounting to the outside of the battery housing. That entails a hole in the housing as well as in any jigs used to stack the cells. Therefore the risk of a short circuit between the cell potential and battery housing is increased by the electrically conductive setup and has to be dealt with. Also the possible cells to be triggered may be limited to where the penetration mechanism can be attached to the battery housing.

The last trigger mechanism, overheating, can be found directly within the SAE J 2464 Sec. 4.4.5's "Passive Propagation Resistance Test" [SAE-International (2009)]. The cell to trigger is being heated uniformly up to

400°C in less than 5 minutes. Once the cell shows thermal runaway reactions the heating device should be switched off. The cell can be heated, for example, with a heat plate built into the cell stack. If no thermal runaway occurs this method can be considered as unfit. An advantage of this method is that the cell encapsulation is not damaged as it would be through nail penetration. Also the induced amount of energy in the system can be limited by the fast heat up and triggering of the cell. A disadvantage is the debris inserted in the system which can cause changes in the thermal properties of the system as well as in the physical setup. A heating plate, for example, heats either two adjacent cells at once, or if insulated to one side, represents a thermal barrier to thermal propagation. A reasonable solution would be replacing a cell with a heating plate that has the exact same physical dimensions and delivering a thermal heat up ramp similar to that of a cell undergoing thermal runaway, which could be determined in advance with an accelerating rate calorimeter (ARC).

Apart from the trigger method, the position of the triggered cell should represent a worst case scenario. However, determining the worst case position is a subject yet to be discussed. So far thoughts go in the following direction: the further a triggered cell is located inside a battery, away from any heat dissipation parts (or heat sinks), and the closer it is to many other cells, the more severe the propagation will be. Also, the distance to any high potential current rails or other critical electronics or components should be considered while choosing the trigger cell position. Another consideration is the trigger method and the necessary space it requires; this can also limit any possible trigger positions.

The chosen trigger method and position should be described as detailed as possible in the test report, so as to enable evaluation and recreation of the test. If no trigger method mentioned leads to a thermal runaway the cell can be considered as safe.

8.3.3 *Measurement equipment and methods*

Both test descriptions [SAE-International (2009); Orendorf et al. (2017)] suggest the following data to be measured before, during and after the test execution:

- Temperature
- Voltage
- Video recording
- Notation of released flames/projectiles

- Pictures before and after test
- Mass of the DUT before and after
- Chemical analysis of gas and its flammability

The relevance of this measured data and its measurement challenges shall now be further discussed.

Temperature over time is probably the most important physical quantity to be recorded. It not only shows maximum temperatures but also can help to further evaluate the paths of heat flux, and based on that, improve future battery designs. Thermo couples (e.g., Typ-K NiCr-Ni) have proven to be feasible and best suitable for temperature measurements during thermal propagation experiments. With the right insulation, like ceramic fibers, thermocouples offer a high but necessary temperature range of up to $T_{\mathrm{maxTypeK}} = 1300°C$ [IEC (2014)]. This is necessary as previous tests temperatures of around $T_{\mathrm{interCellmax}} \approx 950°C$ have been observed between two pouch cells. To obtain a detailed view of the temperature development during thermal propagation, it is necessary to place several thermocouples at different locations on and within the DUT. The number of thermocouples used can thereby be limited by the available measurement equipment and the available space within the DUT.

Figure 8.3 gives an overview of the sensor positioning problem. Depending on the cell format there can either be plenty of space between the cells or nearly no space at all. Figure 8.3 (a) and (b) demonstrates this circumstance. While there is space for sensors between the cylindric cells due to their physical shape, prismatic or pouch cells do not necessarily offer that. Figure 8.3 (c) to (f) each show cross sections of two prismatic or pouch cells with an interjacent sensor. In Fig. 8.3(c), if the sensor is put between

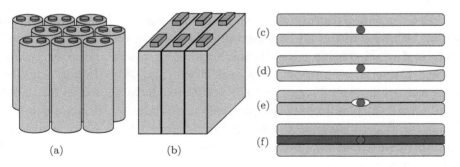

Fig. 8.3 Problems with positioning thermocouples. Space difference between (a) cylindrical cells and (b) prismatic or pouch cells, (c) and (d) change in thermal behavior through thermocouples and (e) possible cell damage, as well as (f) possible solutions.

the two cells, an air gap would change the thermal coupling to the two cells. Figure 8.3(d) and (e) show the two cells pushed together with a jig frame, for example. Hereby, hardcase cells can bend as shown in Fig. 8.3(d) and therefore still be separated by an air gap. In the case of pouch cells, the sensor can impinge the cell, deleting the air gap but causing possible severe damage to the inner cell layers. This case is shown in Fig. 8.3(e). All three cases (Fig. 8.3(c–e)) are not desirable. A possible solution is to embed the sensor in an either existing or additional layer, as seen in Fig. 8.3(f). Existing layers can be heat sink sheets or spacers between the cells, while additional layers should be made out of thermally highly conductive materials (e.g., copper) to maintain the original thermal properties as much as possible. The problem of sensors between cells scales with the actual sensor size, the thinner the sensor the smaller the gap problem.

To obtain a detailed look on the propagation of the thermal runaway perpendicular to each cell, there needs to be one sensor for every 1–2 cells. Additional sensors can give more details about the propagation in other dimensions of the cell module or battery.

Voltage measurements during the thermal propagation tests can give an overview of the propagation progress of the runaway front and can indicate failures in the electrical path, such as short circuits. While a cell module or battery voltage measurement is easy to realize, a direct cell voltage measurement can give additional information about which and when cells are affected. However, a cell voltage measurement also requires additional wires and therefore poses a higher risk of influencing thermal or physical properties of the DUT, and causing potential short circuits due to melting measurement wire insulation. Depending on the number of cells in a DUT a cell voltage measurement of single cells may not be feasible anymore.

Video recording during the test execution delivers information on the venting behavior of the DUT, especially with regard to questions regarding where and when venting gas exits the housing, or where and when ruptures happen. Cameras with frame rates of at least $f_{video} \geq 60\,Hz$ and a full high definition resolution of $r_{video} \geq 1920 \times 1080\,pxl$ have shown to be suitable for this task. Depending on the physical size of the DUT, a total number of $n_{cam} \approx 3 \leq 10$ cameras recording from different angles and positions is recommended. Creating a mixture of close-up views on details (for example, venting openings or plug sockets) and wide angle views of the complete system is useful for gathering as much information as possible. Depending on the severity of a thermal propagation event, close-up cameras might be destroyed if not protected. This should be considered when choosing the type of camera and/or protection device.

Notation of released flames/projectiles during the test or by reviewing the video footage helps to give a quick overview of the course of the experiment in a report without having to go through all data or video footage again. These notes can help a reader put certain temperature or voltage diagrams in context with distinctive events like explosions or sparks.

Pictures of the DUT and surrounding test environment are again important for a later report and can help to evaluate initially inexplicable outcomes or events during the test. While pictures before the execution of the test are mainly to document the initial test setup and conditions of the test environment and DUT, pictures after the test and during the appraisal of the DUT help to evaluate the outcome and possible destruction and failures during the test. In general, pictures deliver an easier way for the reader to judge an experiment without actually being there in person.

Mass determination of the DUT before and after the test is needed to quantify the mass loss during thermal propagation. This mass loss results from the reaction of electrolyte to venting gas leaving the battery housing during the test and the ejection of solid or liquid materials. This mass loss is an indicator for the severity of thermal propagation and will help to classify the outcome. Further explanation will be given in Sec. 8.3.5.

Chemical analysis of the gas and its flammability is a suggested measurement in [Orendorf et al. (2017)]. Experience has shown that this is extremely difficult to do during the propagation test due to possibly rough environmental conditions such as high temperatures, fire, electric arcs and explosions. A gas analysis executed on a single cell with the help of an autoclave, additionally and independent from the thermal propagation test, has proven to be more feasible to determine the gas composition and its amount.

8.3.4 *Experiment setup and conditions*

The design of a cost, time and material efficient test sequence starts at a low assembly level of the DUT (i.e., couple of cells or a cell module) and works its way up to the highest assembly level (complete battery). Depending on the battery size this can be done in a single test, but it can also be necessary to split it up into several assembly level tests. Such a split helps to breakdown the amount of possible failure effects described in Sec. 8.2.3. For example, a low assembly level with only a couple of cells (and therefore low electrical potentials) reduces the risk of electric arcs, in addition to other thermal propagation effects. Once the impact of thermal propagation on a certain level of assembly is understood, controlled and considered safe for the user and environment, a higher assembly level can be tested.

Test conditions should be as realistic as possible. Ambient temperature during thermal propagation tests is $T_{\text{envir}} = 25°C \pm 5°C$, and the DUT should be preheated to $T_{\text{DUT}} = 55°C$ or the maximum operating temperature and SoC should be $\text{SoC}_{\text{DUT}} = 100\%$ [SAE-International (2009)].

The test should start with the triggering of a thermal runaway within the first cell and end after an hour if no propagation appears. If propagation occurs, the test should end after the DUT completely cools down under the start temperature, and therefore propagation can be considered stopped.

8.3.5 Analyzing the results

After the tests it is necessary to evaluate the outcome. This should always be done considering the expected working environment of the battery, though misuse should also be taken into consideration. The European Council for Automotive Research and Development (EUCAR) has defined a list of hazard levels to help battery manufacturers classify and compare the results of abuse tests.

Table 8.1 shows these so-called EUCAR hazard levels (EHLs) with their description and classification criteria. One of the criteria is the mass loss of the DUT — this is why weighing before and after the test described in Sec. 8.3.3 is essential. If a test result meets the classification criteria of several EHLs, the highest level has to be assumed. Despite classification of the general test outcome, evaluation of the measured temperature and voltage, data helps to understand the heat flux during propagation and gives clues for possible design improvements. Temperature measurement from thermocouples should be critically scrutinized because measurement errors and sensor failures are quite common due to the possible rough conditions during thermal propagation. While sensor failures are easy to distinguish, measurement errors can lead to wrong assumptions and interpretations if not properly identified.

Figure 8.4 shows an example reading of thermocouples inbetween pouch cells recorded during a thermal propagation test. T1 represents how the curve of a thermocouple looks without any sensor failures and measurement errors. The measurement starts at the temperature the DUT was preheated to and then nearly instantaneously rises quickly as soon as the thermal propagation

Table 8.1 Description of the modified European Council for Automotive Research & Development (EUCAR) hazard levels [Orendorf et al. (2017)].

Hazard level	Description	Classification criteria and effect
0	No effect	No effect. No loss of functionality.
1	Passive protection activated	No damage or hazard; reversible loss of function. Replacement or re-setting of protection device is sufficient to restore normal functionality.
2	Defect/damage	No hazard but damage to RESS; irreversible loss of function.
3	Minor leakage or minor vent	Visual or audible evidence of leaking or venting. Leak without significant pooling or collection of free liquid. Venting without significant smoke or loss of particulate material. No visual obstruction of the RESS.
4	Major leakage or major vent	Visual evidence of leaking or venting. Leaking with significant pooling or observed free liquid. Venting with significant smoke, solvent vapor, and/or loss of particulate material. Visual obstruction of the RESS by vent gases and/or smoke. Total RESS mass loss < 30%.
5	Rupture	Loss of mechanical integrity of the RESS package, resulting in release of contents. The kinetic energy of released material is not sufficient to cause physical damage external to the RESS. Rupture may be the result of a RESS thermal runaway (but not necessarily). Total RESS mass loss 30–55%.
6	Fire or flame	Ignition and sustained combustion of flammable gas or liquid (1 s sustained fire). Sparks or incandescent material is not considered a fire or a flame.
7	Energetic failure	Fast release of energy sufficient to cause pressure waves (slower than the speed of sound) and/or projectiles that may cause considerable structural and/or bodily damage, depending on the size of the RESS. The kinetic energy of flying debris from the RESS may be sufficient to cause damage as well. Total RESS mass loss $\geq$ 55%.

front reaches the sensor position. The curve shows a smooth progress with no steps or spikes. Sensor curve T2 starts similarly to T1 but shows a clear step at $t = 133\,\text{s}$. This drop in temperature cannot be explained with any physical or chemical reactions during thermal propagation as it would represent a cooling power of roughly

$$P_{\text{cooling}} = \frac{c \cdot m \cdot \Delta T}{\Delta t} \approx \frac{1000\,\frac{\text{J}}{\text{kg}\cdot\text{K}} \cdot 1\,\text{kg} \cdot 500\,\text{K}}{1\,\text{s}} = 500\,\text{kW}, \tag{8.2}$$

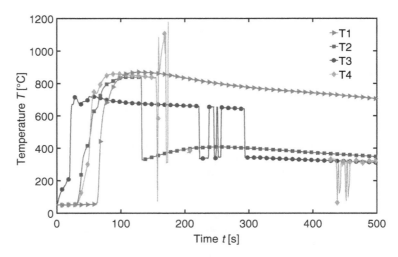

Fig. 8.4 Example of different thermocouple readings and possible sensor errors or failures.

with the specific heat capacity c, the mass of the measured area m, the temperature drop ΔT and the time span of the drop Δt. A more likely scenario is a measurement error. Thermocouples measure the temperature at the point where the two wires connect closest to the measurement device. Here the two wires are short-circuited by venting gas, spilled electrolyte or ash particles as the insulation of the thermocouple is only a netting of ceramic fibers and therefore permeable to liquids and small particles. This leads to a temperature measurement at a point with much lower temperature. T3 shows a similar behavior in the area of $t = 200$ s and $t = 300$ s. Here the short-circuit is not permanent but rather a defective contact that led to signal jumping between the two temperatures. T4 exhibits sharp spikes from $t = 150$ s to $t = 200$ s. Again, these artifacts do not represent real temperature developments within the DUT as it would require very high heating and cooling power analog like the one shown in Eq. (8.2). In this case a short circuit with other thermocouples led to these spikes in temperature readings, caused by the effect described above.

While interpreting temperature readings from thermocouples, the mentioned misreadings and sensor failures always should be considered and any interpretations should be justified and questioned against physical principles. Thermocouples with a tight insulation like stainless steel tubes are less likely to show these artifacts, but are also more expensive and not as mechanically resilient.

8.4 Influencing Variables

There are several tunable variables that influence the system behavior and response to thermal runaway and propagation. This section will discuss some of these influencing variables to give a guideline on how to manipulate the thermal properties of a battery towards a higher resistivity to thermal propagation and therefore towards a safer battery design.

8.4.1 *Cell format*

The physical shape and size of the cell has one of the biggest influences on thermal propagation, tuning the battery behavior at a very low system level. Figure 8.3(a) and (b) demonstrate the different thermal coupling between cells depending on the cell format. While cylindrical cells maximally touch each other on a line contact, prismatic and pouch cells are connected over an area contact, usually including the biggest cell sides. The smaller the contact area between cells, the lower their thermal coupling and therefore the higher their resistivity to thermal propagation, neglecting any coupling over additional parts.

Besides shape, physical cell size is a second factor to consider for thermal propagation resistivity. The trigger mechanisms described in Sec. 8.3.2 always trigger a single cell, as cell failure never occurs simultaneously at several cells.

Figure 8.5 illustrates the advantage of smaller cells to bigger cells. On the left side of the diagram, two prismatic cells with a certain capacity are shown; the front one marked with an (X) is experiencing a failure and is undergoing

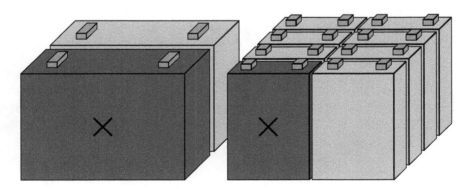

Fig. 8.5 Influence of the cell format on the set free energy during thermal runaway and its transfer to adjacent cells.

thermal runaway. On the right side, eight prismatic cells are drawn, each with a quarter of the capacity of the cells on the left and therefore possessing the same system capacity. Again, one of the front cells marked with an (X) is undergoing thermal runaway. Equation (8.1) describes the relationship between electrochemically stored energy to thermally released energy during a thermal runaway. With the damaged cell on the right only having a quarter of the capacity of the damaged cell on the left, the thermal energy set free is only a quarter on the right. Additionally, the freed thermal energy on the right can be distributed over three cells while the energy on the left gets transferred completely to the second cell, neglecting any heat transfers to the environment. This effect can be scaled by splitting up the cells into six, eight or even more smaller cells.

However, a disadvantage to this strategy is a decrease in energy density due to a small overhead in volume because of the additional cell encapsulation, taps and connection effort.

8.4.2 Energy density

Different cell chemistries and active materials result in different energy densities of lithium-ion cells [Thackeray et al. (2012)]. Even though a steady urge to higher energy densities is observed, this trend is counterproductive to thermal runaway safety, as the thermal stability scales with energy [Doughty (2012)]. Lower onset temperatures for thermal runaway reactions with higher energy densities affect the thermal runaway behavior of a battery; the energy density itself has an impact on the system behavior in case of thermal runaway and propagation. The increase in temperature a cell experiences during its own thermal runaway can be calculated through

$$\Delta T = \frac{E_{\text{thermal}}}{c_{\text{cell}} \cdot m_{\text{cell}}}, \qquad (8.3)$$

with ΔT being the temperature rise of the cell, E_{thermal} the introduced thermal energy, c_{cell} the specific heat of a lithium-ion cell and m_{cell} the mass of a cell. Together with Eq. (8.1), a term to describe the relation between adiabatic temperature rise and energy density can be described as

$$\Delta T \approx \frac{E_{\text{electrical}}}{c_{\text{cell}} \cdot m_{\text{cell}}} = \frac{1}{c_{\text{cell}}} \cdot \omega_{\text{grav}}, \qquad (8.4)$$

with $E_{\text{electrical}}$ being the electrochemically stored energy and the gravimetric energy density ω_{grav}. The gravimetric energy density is hereby approximately proportional to the adiabatic temperature rise of the cell during thermal runaway.

8.4.3 System design

The overall system design of a lithium-ion battery might be the most complex factor to mitigating thermal propagation. It combines a various number of influencing variables, for example, thermal coupling of the cells and cell modules, the path of the venting gas and potential risks of electric arcs.

Figure 8.6 shows an exemplary cross section drawing of two battery designs, one with bad (see Fig. 8.6(a)) and one with good (see Fig. 8.6(b)) thermal propagation mitigation properties. Each of the battery designs has two prismatic or pouch cell modules stacked perpendicular to the drawing plane, one left and one right. Module frames are drawn in green. The left cell module, marked with an (X) is undergoing thermal runaway. The different arrows illustrate the spread of heat (1–3, 7) and venting gas (4, 6) throughout the batteries. In the upper battery design (a), the cells are thermally coupled with the bottom of the battery housing, which leads to heat transfer out of the battery (1), but inevitably also to the second cell module (2). In addition, the coupling through cell frames leads to extra heat being transferred across (3). The lower battery design (b) in comparison has barely any coupling through module frames and only has thermal coupling with the housing on

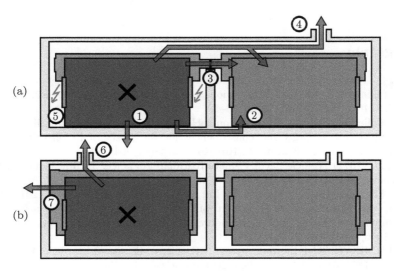

Fig. 8.6 Two cross sections of a battery design, each with two cell modules demonstrating design considerations for thermal propagation mitigation.

the opposite side to the neighboring module (7), thus ensuring excess energy can still leave the system but with a minimum of influence on the other module.

A second design consideration is the venting gas path. In the upper battery design (a), the gas has to stream along the second module to reach a venting opening in the housing (4). Hereby, heat from the gas and carried glowing particles are transferred into the second cell module. In the lower battery design a separate venting opening per module (6) leads to short venting paths within the battery. Additionally, the gas tight separation of the cell modules through the housing protects the second module from any excess gas heat and impurities from soot and ejected active material.

As a final design consideration, the risk of electric arcs should be taken into account. Any uncovered part within the battery with electrical potential can be a source of an electric arc (5). Even if high potentials within the battery are physically far apart from each other, the risk of an electric arc can be given through other conductive materials. The battery housing, for example, if made out of electrically conductive materials can contribute to this risk. Covering and insulating live parts can help reduce the risk of an electric arc but only if it is thermally stable up to high temperatures. Even changes in the physical setup of the battery design during thermal propagation should at least be considered. Melting aluminum heat sinks can lead to newly established electrical connections during thermal propagation and therefore lead to a shift in electrical potential locations.

Bibliography

Biensan, P., Simon, B., Pérès, J., de Guibert, A., Broussely, M., Bodet, J., and Perton, F. (1999). On safety of lithium-ion cells, *J. Power Sources* **81–82**, pp. 906–912, doi:10.1016/S0378-7753(99)00135-4.

Dahn, J., Fuller, E., Obrovac, M., and von Sacken, U. (1994). Thermal stability of Li_xCoO_2, Li_xNiO_2 and λ-MnO_2 and consequences for the safety of Li-ion cells, *Solid State Ionics* **69**, pp. 265–270, doi:10.1016/0167-2738(94)90415-4.

Doughty, D. (2012). Vehicle battery safety roadmap guidance, Technical report, National Renewable Energy Laboratory.

Feng, X., Fang, M., He, X., Ouyang, M., Lu, L., Wang, H., and Zhang, M. (2014). Thermal runaway features of large format prismatic lithium-ion battery using extended volume accelerating rate calurimetry, *J. Power Sources* **255**, pp. 294–301, doi: 10.1016/j.jpowsour.2014.01.005.

IEC (2014). IEC 60584-1 thermocouples — Part 1: EMF specifications and tolerances.

IEC (2016). IEC 62660-2/3 secondary lithium-ion cells for the propulsion of electric road vehicles — Part 2/3.

Kim, G.-H., Pesaran, A., and Spotnitz, R. (2007). A three-dimensional thermal abuse model for lithium-ion cells, *J. Power Sources* **170**, pp. 276–489, doi:10.1016/j.jpowsour.2007.04.018.

Kumai, K., Miyashiro, H., Kobayashi, Y., Takei, K., and Ishikawa, R. (1999). Gas generation mechanism due to electrolyte decomposition in commercial lithium-ion cell, *J. Power Sources* **81–82**, pp. 715–719, doi:10.1016/S0378-7753(98)00234-1.

Lamb, J., Orendorff, C., Steele, L., and Spangler, S. (2015). Failure propagation in multi-cell lithium-ion batteries, *J. Power Sources* **283**, pp. 517–523, doi:10.1016/j.jpowsour.2014.10.081.

Lee, R. (1982). The other electrical hazard: Electric arc blast burns, *IEEE Trans. Ind. Appl.* **IA-18**, pp. 246–251, doi:10.1109/TIA.1982.4504068.

MacNeil, D. and Dahn, J. (2002). The reactions of $Li_{0.5}CoO_2$ with nonaqueous solvents at elevated temperatures, *J. Electrochem. Soc.* **149**, pp. A912–A919, doi:10.1149/1.1483865.

Mikolajczak, C., Kahn, M., White, K., and Long, R. (2011). Lithium-ion batteries hazard and use assessment, Technical report, The Fire Protection Research Foundation.

Orendorf, C., Lamb, J., and Steele, L. (2017). Recommended practices for abuse testing rechargable energy storage systems (RESSs), Technical report, Sandia National Laboratories.

Orendorff, C., Lamb, J., Steele, L., and Spangler, S. (2014). Propagation testing of multi-cell batteries, Technical report, SAND2014-17053, Sandia National Laboratories.

Ponchaut, N., Marr, K., Colella, F., Somadepalli, V., and Horn, Q. (2015). Thermal runaway and safety of large lithium-ion battery systems, *The Battcon 2015 Proc.*, pp. 17.1–17.10, http://www.battcon.com/PapersFinal2015/17%20Ponchaut%20Paper%202015.pdf.

SAE-International (2009). Electric and hybrid electric vehicle rechargeable energy storage system (RESS) safety and abuse testing, http://standards.sae.org/j2464_200911/, standard SAE J2464 NOV2009.

Spotnitz, R. and Franklin, J. (2003). Abuse behavior of high-power, lithium-ion cells, *J. Power Sources* **113**, pp. 81–100, doi:10.1016/S0378-7753(02)00488-3.

Thackeray, M., Wolverton, C., and Isaacs, E. (2012). Electrical energy storage for transportation — Approaching the limits of, and going beyond, lithium-ion batteries, *Energy Environ. Sci.* **5**, pp. 7854–7863, doi:10.1039/C2EE21892E.

Tobishima, S. and Yamaki, J. (1999). A consideration of lithium cell safety, *J. Power Sources* **81–82**, pp. 882–886, doi:10.1016/S0378-7753(98)00240-7.

Underwriters Laboratories Inc. (2012). UL 1642 standard for safety — Lithium batteries, 5th Edition.

Wang, Q., Ping, P., Zhao, X., Chu, G., Sun, J., and Chen, C. (2012). Thermal runaway caused fire and explosion of lithium-ion battery, *J. Power Sources* **208**, pp. 210–224, doi:10.1016/j.jpowsour.2012.02.038.

Chapter 9

Potential of Capacitive Effects in Lithium-ion Cells

Alexander Uwe Schmid and Kai Peter Birke

This chapter intends to provide insights into how capacitance effects are caused in lithium-ion (Li-ion) cells and how to possibly use these short-term effects on the system level. Therefore, fundamentals of capacitive effects in double-layer (DL) capacitors and Li-ion cells are addressed in Secs. 9.1 and 9.2. In order to know the potential of capacitive effects, ways of measurement and modeling are proposed in Sec. 9.3. Subsequently, in Sec. 9.4 an estimation of usable capacitance on the system level is made and the 48 V system for hybrid electric vehicles (HEV) is introduced as a potential application field.

9.1 Brief Introduction to the Principles of Electrostatic and Electrochemical Storage

In order to answer the question of how much capacitance is present in a typical Li-ion cell, it is important to define the term "capacitive effects in Li-ion cells". Two definitions appear logical: (1) Time-limited (short-term) release of charge without considering and distinguishing the physico-chemical mechanisms, (2) Separation of pure double-layer capacitive from electrochemical effects. We want to address both definitions as in fact many authors do. Short-term effects which are not subject to electrostatic principles are therefore called *pseudocapacitive effects*.

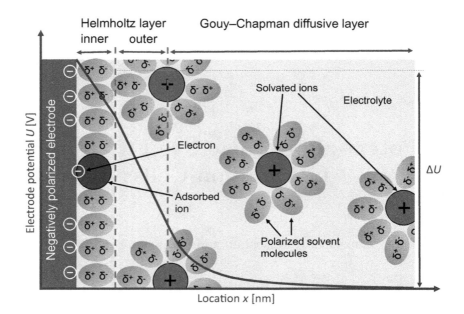

Fig. 9.1 Illustration of the Bockris–Devanthan–Müller model (BDM model). It includes an inner Helmholtz layer (the solvent has influence on the electric field due to possible adsorption of ions, alignment of the dipoles, and radii of the solvent molecules), an outer Helmholtz layer where the radii of the solvated ions define the thickness of this layer, followed by a diffusive layer observed by Louis Georges Gouy and David Leonard Chapman in 1913. Aligned ions and dipoles generate drop in potential ΔU.

9.1.1 Double-layer capacitance

In general, a DL capacitance can be defined as a charge accumulation at electrode interfaces where ionic and electronic charges are separated [Conway et al. (1997)]. The electrolyte containing conducting salt dissociates to negatively (anions) and positively (cations) charged ions. The dipoles of the solvent (part of the electrolyte) are aligned, attracted and form an ordered structure around the ions, as seen in Fig. 9.1. When a voltage is applied to a DL capacitance, electrical dipole moments of the solvents and ionic charge force the ion to move to the corresponding polarized electrode. The DL capacitance is

$$C_{\mathrm{dl}} = \Delta Q / \Delta U, \tag{9.1}$$

with ΔU as the potential difference across the interface caused by accumulated charges ΔQ on the interface [Conway et al. (1997)].

A capacitor able to store the complete energy via this DL is mentioned here in the following *double-layer capacitor* (DLC). In contrast to electrochemical energy storage where the energy is stored mainly by Faradaic charge transfer, DLCs store energy by an electrostatic field.[1]

The capacitance of an idealized plate capacitor can be calculated with

$$C_{\text{plate}} = \epsilon_0 \epsilon_r A/d, \tag{9.2}$$

where ϵ_0 is the vacuum permittivity, ϵ_r is the relative permittivity of a dielectric medium, A is the planar electrode area and d is the distance between the two electrode plates. Although the mechanism of ionic charge separation of a DLC is different from a simple plate capacitor (in contrast to foil capacitors, a DLC has a double-layer consisting of ionic and electronic charge layers), the principle of optimization is still the same: it is easy to see from Eq. (9.2) that adjusting ϵ_r, A or d changes the capacitance. This means that the electrode interface area S,[2] and thickness of DL with less than 1 nm [Plett (2015); Amatucci et al. (2001)] are essential for high capacitance values and the amount of energy that can be stored, since ϵ_r is limited by the specific material properties. Large specific capacitance values of about $C_{\text{dl}}/m = 25\,Fg^{-1}$ can be achieved for DLCs by means of carbon powders, which have high specific surface areas $(S/m = 100 - 2000\,m^2 g^{-1})$ [Conway et al. (1997)].

The Bockris–Devanthan–Müller model (BDM model), schematically illustrated in Fig. 9.1, can give an understanding about influencing the selection of electrolyte and its solvents on DL capacitance, since the DL variates with size of the ions and concentration of electrolyte [Zhang et al. (2009)]. The model contains an inner and outer Helmholtz layer, a Gouy–Chapman diffusive layer and specific adsorption on the electrodes' surface. In case of adsorption, the solvated ion has to strip off its solvent molecules and enter into a RedOx reaction with the electron. Strictly speaking, a charge transfer can also occur on a DL. By our definition of the DL capacitance, only the electrostatic effects are understood as DL capacitance.

The following Eq. (9.3) describes the thickness of the Helmholtz outer layer κ^{-1} [Amatucci et al. (2001)],

$$\kappa^{-1} = \left(\frac{\epsilon_r \epsilon_0 k_{\text{B}} T}{F^2 \sum N_{\text{i}} z_{\text{i}}^2} \right)^{1/2}, \tag{9.3}$$

---

[1] Energy density E/m of a nonaqueous DLC is about 2–5 Wh/kg [Amatucci et al. (2001)] compared to Li-ion cells with 100–250 Wh/kg.
[2] Here, the variable A stands for planar electrode area [m²], whereas S [m²] is the effective surface area of specific material that is available to the ions for charge accumulation.

where N_i and z_i are the concentration and the charge number of an ion, T is the temperature, k_B is the Boltzmann constant and F is the Faraday constant. The outer Helmholtz layer κ^{-1} of monovalent cations is only about 10 Å thin [Amatucci et al. (2001); Zhang et al. (2009)].

9.1.2 Intercalation

During intercalation, ions are stored on interstitials. A typical example for intercalation processes is a Li-ion cell: Both electrodes usually work with intercalation: metal oxides on the positive and graphite on the negative side. The so-called *ion swing* ensures that the ions are de-/intercalated from one side to the other without elementally changing the host structure of the electrode. Li-ion cells store energy mainly by intercalation of Li-cations with a corequisite electron transfer. A charge transfer of a Li-ion (Li$^+$) can be written according to Ong and Newman (1999) as:

$$\mathrm{Li}^+ + \mathrm{e}^- + \Theta_s \rightleftarrows \mathrm{Li} - \Theta_s. \tag{9.4}$$

The chemical reaction (9.4) looks like a RedOx reaction, where a Li-ion reacts to Li-metal, but where in fact the cation occupies a free site Θ_s of the intercalation material. At the same time an electron migrates into the host lattice.

In electrochemistry, reversible and irreversible reactions are potential-dependent, where only a certain small voltage range is valid for operating the electrochemical cells, with respect to safety, long lifetime and energy density $E/m \propto U_{\mathrm{cell}}$. The cell voltage U_{cell} is the potential of the positive electrode versus reference potential ϕ_{pe}, minus the potential of the negative electrode ϕ_{ne} versus reference potential $U_{\mathrm{cell}} = \phi_{\mathrm{pe}} - \phi_{\mathrm{ne}}$.

The general thermodynamic dependency of cell potential U_{cell} in terms of occupancy of three-dimensional lattice sites is [Conway et al. (1997)]

$$U_{\mathrm{cell}} = U_0 + (\mathrm{R}T/z\mathrm{F}) \cdot ln(\Theta/(1-\Theta)). \tag{9.5}$$

The variable U_0 is the cell potential under equilibrium conditions, Θ refers to occupied lattice sites, $1-\Theta$ are the corresponding unoccupied sites, R is the gas constant and z is the number of oxidized or reduced electrons per molecule.

9.1.3 Pseudocapacitance

Conway et al. (1997) define a pseudocapacitance as a process where the Faradaic charge is transferred through a phase boundary in which the charge transfer depends on potential. The authors regard the dependency of potential and the charge transfer as two distinguishing features between pseudo- and DL capacitance. Conway et al. (1997) mention three categories as cause for pseudocapacitive effects:

- Adsorption of an electroactive species at electrode interfaces
- RedOx processes on functional groups on the electrode surface
- Surface near intercalation/deintercalation

Supercapacitors, sometimes also called *pseudocapacitors*, use the principle of the Faradaic charge transfer via adsorption as energy storage. This hybrid type of capacitor combines the ability to store charge on both a non-Faradaic (electrostatic via the DL) and a Faradaic way. They often consist of metal oxide or polymer at one or both electrode sides [Plett (2015)].

Focusing on Li-ion cells, intercalation electrodes can behave like a combination of a pure battery (pure RedOx reaction) and a supercapacitor [Conway et al. (1997)]. Conway defines pseudocapacitance in Eq. (9.6).

$$C_\phi = d(\Delta Q)/dU. \tag{9.6}$$

Bruce and Saidi (1992) refer the pseudocapacitive effect to adsorption and intercalation. Their *adatom model* describes the mechanism of electrochemical intercalation of Li-ions into titanium disulphide TiS_2 as a sequential process of adsorption and lattice incorporation. When a Li^+ intercalates into the specific electrode, the ion first strips off its solvent molecules, then it adsorbs onto the surface of the electrode before it incorporates. Immediately an electron flows into the electrode to retain electroneutrality [Bruce and Saidi (1992)]. This adsorption process is modeled in Fig. 9.6 in Sec. 9.3.1.1.

9.2 Similarities and Differences between Capacitors and Lithium-ion Cells

The schematic drawing in Fig. 9.2 of a Li-ion cell addresses important similarities to a DLC, in particular the alignment of the ions through the cell voltage U_{cell}. The construction of a Li-ion cell is very similar to a DLC, as the following points indicate:

- General structure: Current collectors, electrodes, electrolyte, separator.
- Electrode materials: Carbon as a conductive additive in electrodes and graphite as the negative electrode.
- Electrolyte: Ion conducting medium, non-conducting for electrons. Dissociated and polarized ions causing DL capacitance.

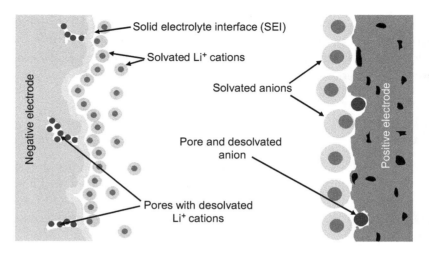

Fig. 9.2 Analogies of the principle of energy storage of a Li-ion cell to the BDM model: (1) Alignment of dissolved and solvated cations and anions, (2) Solid electrolyte interface (SEI) as a kind of inner Helmholtz layer, and (3) Adsorbed and desolvated ions in the pores. Additionally to adsorbed ions, the electrode lattice with occupied sites near to the electrode surface.

Several questions arise from the fact that every Li-ion cell has a superposed DL capacitance: Which influence has (1) cell material and (2) the solid electrolyte interface (SEI) on the capacitance? (3) How to measure the capacitance? (4) How big is the DL capacitance in Li-ion cells? (5) How to apply capacitive effects on the system level? The next sub-/sections address these questions.

9.2.1 Carbons as electrode material

These similarities leading to a superposed DL capacitance in Li-ion cells give rise to the question about the influence of cell material. Since carbon is a common material for both DLCs and Li-ion cells, this subchapter describes the characteristics of carbon in its various states.

Carbons combine high electrical conductivity, low-cost and chemical stability [Daniel and Besenhard (2012)], which make them attractive as electrode material or a conductive additive in electrochemical cells and DLCs.

DLCs are predestined for high-power applications because of the quick alignment of ions in the electrolyte after a potential is set. There is a general correlation of high specific DL capacitance C_{dl}/m and surface area S of carbon electrodes [Gryglewicz et al. (2005)]. Amorphous carbons such as *carbon black* or *active carbon* can have high specific surface areas of $S/m > 1000\,\text{m}^2\text{g}^{-1}$, small particle size lower than $p = 50\,\text{nm}$ and less density than graphite with $\delta = 2.25\,\text{gcm}^{-3}$ [Daniel and Besenhard (2012)].

In contrast to amorphous carbons, graphite has three-dimensional ordering of its layers consisting of hexagonal rings (six carbon atoms combined). The weak van der Waals bonds (cohesion) between these layers form the typical graphite structure [Daniel and Besenhard (2012)]. As a result, Li-ions can intercalate without elementally changing the host structure.

Electrode production implies coating the active material onto the current collector. Binder material is used to bond the particles together and to connect the slurry (active material) with the current collector. Good electrical conductivity σ_{con} of the electrode material is essential for the performance of DLCs [Gryglewicz *et al.* (2005); Qu and Shi (1998)] and Li-ion cells. Since the conductivity of metal oxides is relatively small in comparison to graphite,[3] larger amounts of a conductive additive are necessary.

Small particles with a high surface area increases the electrical conductivity. The particle size of graphite in Li-ion cells is ($p = 2 - 40\,\mu\text{m}$, typical $p = 10\,\mu\text{m}$ [Zaghib *et al.* (2000)]) relatively small as compared to other electrode materials [Daniel and Besenhard (2012)].

The possibility of further reducing the particle size of graphite in order to produce even greater surface area is limited in Li-ion cells. The lower the average particle size of graphite, the higher the irreversible capacity loss in Li-ion cells due to electrolyte decomposition at the boundary layer (SEI) [Zaghib *et al.* (2000)].

9.2.2 The solid electrolyte interface

A major characteristic of a Li-ion cell is the solid electrolyte interface (SEI) on the graphite electrode. The carbonate solvents react below $1.5\,\text{V}$ versus Li/Li^+ to surface layers [Levi and Aurbach (1997)]. These surface layers cover the graphite electrode and hence prevent further side reactions between the negative electrode and the electrolyte. The prevailing electrolyte solvent is ethylene carbonate (EC), to suppress further side reactions or co-intercalation [Daniel and Besenhard (2012)]. The SEI as ion conducting layers contributes to cell impedance at certain frequencies. It is visible in the

---

[3] Li_xC_6: $\sigma_{\text{con}} = 0.350\,\text{Scm}^{-1}$ versus $\text{Li}_y\text{Mn}_2\text{O}_4$: $\sigma_{\text{con}} = 8.55 \cdot 10^{-3}\,\text{Scm}^{-1}$ [Ong and Newman (1999)].

electrochemical impedance spectroscopy (EIS)[4] at higher frequencies [Bruce and Saidi (1992)].

The surface area of lithiated graphite Li_xC_n increases strongly compared to non-lithiated graphite [Daniel and Besenhard (2012)]. Since the solvated ion strips off its solvent molecules during intercalation, the SEI is constantly growing over its lifetime, resulting primarily in an increased internal cell resistance. Solvent co-intercalation is another cause for cell aging in Li-ion cells [Vetter *et al.* (2005)]. Volume changes over 100% can occur during co-intercalation for large solvent molecules in lithiated carbons [Daniel and Besenhard (2012)].

The polarization caused by the additional SEI layer and the change in surface area must be considered when focusing on capacitive effects. An analytical method to describe the currently available electrode surface area (and hence to calculate the available capacitance over its lifetime) seems to be very complex. Each factor, such as electrode porosity, particle size of active material, binder and conductive additive, electrolyte composition, current electrode potential and internal cell temperature can accelerate or at least influence the chemical reaction of species leading to a changed surface area.

9.2.3 *Summary*

There are various of types of carbons with respect to porosity and structure used in electrochemical cells. The conductive additives in both electrodes yield to higher surface areas, in this example, especially at the positive metal oxide electrode, for reasons of the required cell conductivity. Regarding the conductive additive carbon in both Li-ion electrodes, it can be assumed that a superposed capacitor is located in the Li-ion cells. Hence, type and composition of the electrode material plays an important role in terms of DL capacitance.

Since DL capacitance strongly depends on the distance between charge separation, surface layers influence the electrochemical reactions between the electrolyte and electrode. The determination of the surface area of the SEI in Li-ion cells is, however, difficult due to a complex layer structure of several chemical products and ongoing reactions (SEI formation and increase of SEI over its lifetime).

---

[4]More information about the EIS in Sec. 9.3.1.

9.3 Methods of Measurement of Capacitive Effects

9.3.1 *Electrochemical impedance spectroscopy*

An AC impedance measurement technique is *electrochemical impedance spectroscopy* (EIS), in which a response signal is analyzed using a known sinusoidal excitation signal. Regarding galvanostatic mode with excitation current

$$i(\omega, t) = \hat{i} \sin \omega t + \phi_i, \qquad (9.7)$$

where ϕ_i is the phase angle of the current, $\hat{i}$ is the amplitude, t is the time and ω is the angular frequency. The ratio of excitation to response signal from Eq. (9.8) is used to calculate the complex impedance $\underline{Z}(\omega)$ shown in Eq. (9.9).

$$u(\omega, t) = \hat{u} \sin \omega t + \phi_u, \qquad (9.8)$$

$$\underline{Z}(\omega) = \frac{i(\omega, t)}{u(\omega, t)} = \frac{\hat{i}}{\hat{u}} \cdot e^{j(\phi_i - \phi_u)} = \underline{Z}'(\omega) + j\underline{Z}''(\omega). \qquad (9.9)$$

Analogously to Eq. (9.7), $\hat{u}$ and ϕ_u in Eq. (9.8) are the amplitude and the phase angle of voltage response u. The complex impedance can be divided into a real part $\underline{Z}'(\omega)$ without a phase shift between current and voltage $\Delta\phi = \phi_i - \phi_u = 0°$ and an imaginary part $\underline{Z}''(\omega)$ where $|\Delta\phi| > 0°$ applies. Three conditions have to be met for EIS to provide interpretable measurement results: (1) The excitation signal must be small enough to feed a linear response signal, (2) The system must retain its initial state after measurement, and (3) The response signal is only caused by the excitation signal [Gamry Instruments (2017)]. Then, EIS provides information about the complex impedance $\underline{Z}(\omega)$ over a certain frequency range.[5] The interpretation of $\underline{Z}'(\omega)$ and $\underline{Z}''(\omega)$, in terms of equivalent circuit elements, for instance, depends on the model to be used.

Figure 9.3 shows a typical Nyquist plot of a Li-ion cell (the imaginary part $Z''(\omega)$ plotted against the real part $Z'(\omega)$).[6] Electrical circuit models (ECMs) like the Randles circuit can help to interpret the data measured by EIS. For example in Fig. 9.3, the internal resistance of a cell is represented by R_i, C_{dl} stands for the DL capacitance, R_{ct} is the charge transfer resistance,

---

[5] Usually $100\,\text{kHz} \geq f \geq 10\,\text{mHz}$ in practice.
[6] In this case, a Randles circuit [Randles (1947)] generated the characteristic impedance curve with appropriate parametrized values for the elements.

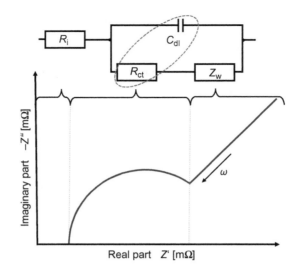

Fig. 9.3 Frequency-dependent impedance $\underline{Z}(\omega)$ of a Randles circuit [Randles (1947)]. The resistance R_i generates the offset in Z'. R_{ct} and C_{dl} in parallel cause the semicircle in the Nyquist plot. The branch at lower frequencies is caused by the Warburg impedance $\underline{Z}_w$.

and Z_w is the Warburg impedance which describes the impedance of diffusion controlled mass transfer in a cell [Bard and Faulkner (1980)]. The half circle in Fig. 9.3 is caused by C_{dl} and R_{ct}. The occurrence of a second semicircle at higher frequencies (not visible in Fig. 9.3) is either traced back to the SEI (see Sec. 9.2.2) or to particle-to-particle resistance [Choi and Pyun (1997)] of the electrode material.

9.3.1.1 Modeling approaches based on equivalent circuit elements

Many ECMs of electrochemical cells include several ideal and/or non-ideal capacitances, which model the special effects associated with the physical cause. We show four modeling approaches in the following, each adjusted and fitted to a special cell configuration.

Modeling of a supercapacitor

Taberna *et al.* (2003) present an approach to interpret a supercapacitor as a resistance and a capacitance in series, both depending on frequency. The impedance of an ideal capacitor can be described by following equation:

$$Z(\omega) = \frac{1}{j\omega C}. \tag{9.10}$$

Insertion of Eq. (9.9) in Eq. (9.10) (considering a frequency-dependent capacitance $C(\omega)$) leads to Eqs. (9.11) to (9.13).

$$C(\omega) = C'(\omega) - jC''(\omega), \qquad (9.11)$$

$$C'(\omega) = \frac{-Z''(\omega)}{\omega |Z(\omega)|^2}, \qquad (9.12)$$

$$C''(\omega) = \frac{Z'(\omega)}{\omega |Z(\omega)|^2}. \qquad (9.13)$$

The real part of the capacitance $C'(\omega)$ is proportional to the susceptance (imaginary part of complex admittance). The imaginary part $C''(\omega)$ is the conductance divided by ω. Since a resistance and a capacitance in series is only a rough model for Li-ion cells, analytical transformation of measured impedances $Z'(\omega)$ and $Z''(\omega)$ can become more complex.

Analytical transformation of measured impedance to a Randles circuit and interpretation

A second Randles circuit is shown in Fig. 9.4 on the left side. This ECM can be transformed to the ECM on the right side in Fig. 9.4. Bard and Faulkner (1980) summarize the elements R_{ct} and $\underline{Z}_w$ to a Faradaic impedance $\underline{Z}_f$, which strongly changes with frequency in contrast to $\underline{C}_{dl}$ and R_i. Thus the electrical circuit elements of $\underline{Z}_f$ are only valid for the frequency that has been used for parametrization. In the same way, $\underline{Z}_f$ can be transformed to a series connection of resistance R_s and pseudocapacity $\underline{C}_s$ [Bard and Faulkner (1980)]. If the Faradaic current i_f, which flows through Z_f, was measurable, the derivation of voltage drop over both elements (R_s and C_s) would be as in Eq. (9.14).

$$\frac{dU}{dt} = R_s \frac{di_f}{dt} + \frac{i_f}{C_s}. \qquad (9.14)$$

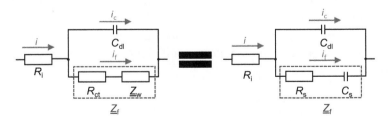

Fig. 9.4 Two ECM variants of the same electrochemical cell. Retraced figure according to Bard and Faulkner (1980), p. 323.

Then it follows for a sinusoidal current as excitation signal (EIS measurement) $i_f = \hat{i}\sin\omega t$:

$$\frac{dU}{dt} = R_s \hat{i}\omega \cos\omega t + \frac{\hat{i}}{C_s}\sin\omega t. \tag{9.15}$$

Equation (9.16) shows the voltage dependency for a chemical RedOx system where U depends on current i and the soluble concentrations C_O, C_R of species [Ox] and [Red] in the electrolyte with zero distance to electrode surface.

$$\frac{dU}{dt} = R_{ct}\frac{di}{dt} + \beta_O\frac{dC_O}{dt} + \beta_R\frac{dC_R}{dt}. \tag{9.16}$$

The following applies for Eq. (9.16):

$$R_{ct} = [\delta U/\delta i]_{C_O, C_R}, \tag{9.17}$$

$$\beta_O = [\delta U/\delta C_O]_{i, C_R}, \tag{9.18}$$

$$\beta_R = [\delta U/\delta C_R]_{i, C_O}. \tag{9.19}$$

Equation (9.20) is derived by Bard and Faulkner (1980) from Eq. (9.16) under the condition of steady states[7] and semi-infinitive linear diffusion[8]:

$$\frac{dU}{dt} = \left(R_{ct} + \frac{\sigma}{\omega^{1/2}}\right)\hat{i}\omega\cos(\omega t) + \hat{i}\sigma\omega^{1/2}\sin\omega t. \tag{9.20}$$

With the mass transfer coefficient σ:

$$\sigma = \frac{1}{zFS\sqrt{2}}\left(\frac{\beta_O}{D_O^{1/2}} - \frac{\beta_R}{D_R^{1/2}}\right). \tag{9.21}$$

In general, mass transfer includes migration (movement of particles caused by an electric field), diffusion (movement of particles caused by a chemical gradient) and convection (fluid flow) [Bard and Faulkner (1980)]. The mass transfer coefficient σ accounts for the change in concentration and concentration ratios with the diffusion coefficients D_O and D_R of the species [Ox] and [Red].

---

[7] Constant surface concentrations before and after impedance measurement: Concentrations $C_O(x, t=0) = C_O^*$, $C_R(x, t=0) = C_R^*$. (C_O^*, C_R^*: concentrations in the bulk).
[8] Only distance x to the electrode surface influences the concentration gradients $C_O(x,t)$, $C_R(x,t)$. More information in [Bard and Faulkner (1980)] on page 252.

Then, Bard and Faulkner (1980) equate the coefficients of Eqs. (9.15) and (9.20) to identify R_s and C_s:

$$C_s = 1/(\sigma \omega^{1/2}), \qquad (9.22)$$

$$R_s = R_{ct} + \sigma/\omega^{1/2}. \qquad (9.23)$$

The mass transfer terms in Eqs. (9.22) and (9.23) are the real and imaginary part of the Warburg impedance Z_w:

$$R_w = \sigma/\omega^{1/2}, \qquad (9.24)$$

$$C_w = 1/(\sigma \omega^{1/2}), \qquad (9.25)$$

$$X_w = 1/(\omega C_w) = \sigma/\omega^{1/2}, \qquad (9.26)$$

$$Z_w = (R_w^2 + X_w^2)^{1/2} = \sigma \cdot (2/\omega)^{1/2}. \qquad (9.27)$$

The phase angle of Z_w with $\phi_w = 45°$ can be identified in the Nyquist plot at low frequencies, where solid diffusion determines the response signal referring to Li-ion cells [Plett (2015)].

The linearity of amplitudes of the response signal to the excitation signal is one precondition for EIS measurement. With that, linearity follows according to [Bard and Faulkner (1980)]:

$$R_{ct} = \frac{RT}{zFi_0}, \qquad (9.28)$$

$$\sigma = \frac{RT}{z^2 F^2 S \sqrt{2}} \left(\frac{1}{D_O^{1/2} C_O^*} + \frac{1}{D_R^{1/2} C_R^*} \right). \qquad (9.29)$$

The variable i_0 is the exchange current. C_O^* and C_R^* are the bulk concentrations far away from the electrode surface. The mass transfer coefficient σ variates with the diffusion coefficients, the concentration of bulk species at initial conditions, the temperature and the surface area.

Based on the model in Fig. 9.4, the measured quantities $Z'(\omega)$ and $Z''(\omega)$ can be expressed as quantities of the equivalent circuit elements R_i, R_{ct}, C_{dl} and σ [Bard and Faulkner (1980); Sluyters-Rehbach and Sluyters (1970)]:

$$Z'(\omega) = R_i + \frac{R_{ct} + \sigma \omega^{-1/2}}{(C_{dl} \sigma \omega^{1/2} + 1)^2 + \omega^2 C_{dl}^2 (R_{ct} + \sigma \omega^{-1/2})^2}, \qquad (9.30)$$

$$Z''(\omega) = \frac{\omega C_{dl}(R_{ct} + \sigma \omega^{-1/2})^2 + \sigma \omega^{-1/2}(\omega^{1/2} C_{dl} \sigma + 1)}{(C_{dl} \sigma \omega^{1/2} + 1)^2 + \omega^2 C_{dl}^2 (R_{ct} + \sigma \omega^{-1/2})^2}. \qquad (9.31)$$

Conway et al. (1997) show analogies between RedOx and intercalation systems (see Eqs. (9.32) and (9.5)) giving rise to pseudocapacitive effects. The potential U of a RedOx system can be described by means of the Nernst equation with concentrations of species [Ox] and [Red] as C_O and C_R, approximated as activities. According to Conway et al., the potential of pseudocapacitive effects depends on the relation of free energy to $ln\Theta/(1-\Theta)$.

$$U = U_0 + (RT/zF) ln(C_O/C_R), \qquad (9.32)$$

$$U = U_0 + (RT/zF) ln(\Theta/(1-\Theta)). \qquad \text{(see 9.5)}$$

The analogies of Eqs. (9.5) and (9.32) can lead to a new interpretation of $\beta^* = [\delta U/\delta\Theta]_i$, comparing the interpretations of pseudocapacitance C_s from [Bard and Faulkner (1980)] (see Eqs. (9.22) and (9.21)) and Conway's pseudocapacitive effects in Li-ion cells. The state of free lattice sites in an intercalation electrode influences the cell voltage, as β_R, β_O and σ do in the case of a RedOx system (see Eqs. (9.18), (9.19) and (9.21)). The typical $U - SoC$[9] behavior of a Li-ion cell is presented in Fig. 9.5. The solid line shows the cell voltage calculated using Eq. (9.5).[10] The dashed lines present the relations of SoC · U (Gibbs energy $G = QU$ with Q as charge capacity) to $ln\Theta/(1-\Theta)$ [Conway et al. (1997)]. This relation (SoC · U)/($ln\Theta/(1-\Theta)$) becomes small at very high and very low SoC. The previously defined variable β^* and thus the resulting mass transfer coefficient σ becomes large, considering high or low SoC.[11] Reciprocally, the pseudocapacitance could be increased when a large number of free lattice sites are available. Due to different electrode materials in Li-ion cells (metal oxides and graphite, for instance) with their respective diffusion coefficients, electrode porosity and surface area, the ratio and hence the shape of the curves in Fig. 9.5 can change accordingly.

Adsorption model (adatom model)

Figure 9.6 shows an additional way to interpret the alternating voltage-current behavior of a Li-ion cell. This model from Bruce and Saidi (1992) combines a Randles circuit, as shown in Fig. 9.4, with an additional parallel circuit of a pseudocapacitance C_s, a constant-phase-element $\underline{Z}_{CPE}$ and

---

[9]SoC: State of Charge.
[10]We set occupied lattice sites Θ equal to SoC.
[11]We look at the electrode as a whole, meaning all lattice sites can be occupied. The solid-state diffusion is sufficiently fast.

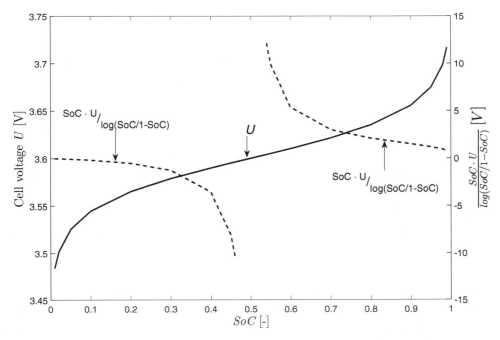

Fig. 9.5 Dependency of cell voltage versus occupied lattice sites (or SoC) calculated with Eq. (9.5). U_0 is set to 3.6 V, T = 293 K, z = 1. The resulting relation of the s-shaped cell voltage times the SoC to $ln\Theta/(1-\Theta)$, shown as dashed lines, is to show the potential of pseudocapacitance [Conway et al. (1997)].

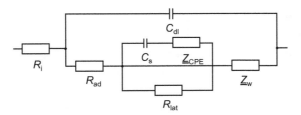

Fig. 9.6 Adatom model of a Li-ion cell. The adsorption process is modeled by additional elements. Retraced figure according to Bruce and Saidi (1992).

a resistance R_{lat}. The resistance R_{ad} explicitly represents the adsorption resistance, whereas R_{lat} stands for the intercalation resistance. The pseudocapacitance C_s is connected here in parallel to R_{lat}, which also means that in this model, the surface near intercalation is responsible for C_s. The adsorption resistance R_{ad} causes an over-voltage, which in turn affects C_{dl}. One challenge facing this theory (further information in Sec. 9.1.3) is the characterization of the effects of adsorption and intercalation, since both

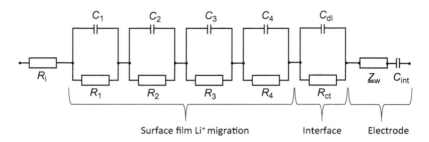

Fig. 9.7 Multi-layer model of the intercalation process of Li$^+$ into graphite, drawn according to Levi and Aurbach (1997). The cell capacity is modeled as capacitance C_{init}.

processes can hardly be separated from another concerning place and time [Bruce and Saidi (1992)].

Multi-layer model

The intercalation process of Li$^+$ into graphite contains (1) Movement of ions in electrolyte solution, (2) Migration through SEI, (3) Charge transfer with electrons, (4) Diffusion in the electrode (solid-state) and (5) Charge accumulation [Levi and Aurbach (1997)]. In particular, the multi-layers of the SEI (at the phase boundary between the electrolyte and graphite) can be modeled using several RC-circuits in series, where each RC-circuit stands for one of those layers [Levi and Aurbach (1997)]. Figure 9.7 shows a way to model the intercalation process of Li$^+$ into graphite. The five RC-circuits model the high-frequency part of the impedance. This model showed excellent fitting results [Levi and Aurbach (1997)] for thin graphite electrodes (under 10 μm), whereas thicker electrodes have a more complex EIS impedance, for instance, an additional electronic resistance [Levi and Aurbach (1997)]. Each RC-circuit has its own time constant. At 0.05 V versus Li/Li$^+$, the fit at high frequencies revealed a DL capacitance of $C_{dl}/A \approx 400\,\mu\text{F cm}^{-2}$ per electrode area[12] with $\tau_{dl} \approx 5.5$ ms of the fifth RC-circuit. The capacitance of a finite-length Warburg impedance Z_w, fitted to the observed data is $C_w/A \approx 220\,\text{mF cm}^{-2}$ with $\tau_w = 75$ s. The variable that accounts for charge accumulation or consumption is $C_{init}/A \approx 83\,\text{mF cm}^{-2}$ [Levi and Aurbach (1997)].

Levi et al. (1999) matched the response signal of a thin Li$_{1-x}$CoO$_2$ electrode by almost the same ECM model as shown in Fig. 9.7 (three instead

---

[12] Specific electrode capacitance C/A: Capacitance per electrode area A. The specific capacitance per surface area C/S is smaller.

of four RC-circuits connected in series standing for surface film migration). By interpreting this model based on a thin $Li_{1-x}CoO_2$ electrode, it shows that film layers can exist, even on the surface of cathode materials.

9.3.2 Cyclic voltammetry

The linear sweep cyclic voltammetry (CV) can be used to distinguish between pseudocapacitance and DL capacitance [Conway (1991)]. The current of a pure capacitance corresponds to the differential equation $i = C \cdot dU/dt$. Under the precondition that the DL capacitance C_{dl} is constant with rising voltage U [Conway (1991)], the shape of the i-U-plot must be rectangular for moderate and constant sweep rates $sr = dU/dt$. As a key feature of a DLC, its capacitance $C_{dl} = i/sr$ is relatively invariant for a wide range of sr due to the inherent high current capabilities. Dipole orientation and attachment of the ions to the boundary layer can limit the kinetics of a DLC only for very high sweep rates [Conway (1991)]. Figure 9.8(a) presents an ideal rectangular shape of capacitance over the cell voltage. An exemplary phase transition which can occur during lithium insertion/deinsertion, is shown in Fig. 9.8(b). Levi and Aurbach (1997) link the EIS variable C_{init} (ECM from Fig. 9.7) to results of their CV measurement over the equation $C_{int}(U) = i/sr$.

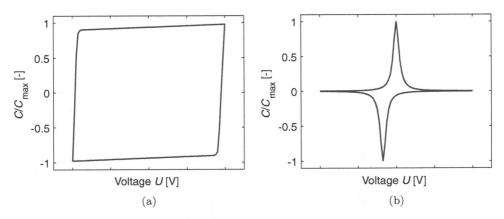

Fig. 9.8 Qualitative shapes of non-Faradaic and Faradaic capacitance. (a) Ideal rectangular shape of a DLC. When the sweep rate changes, the current response changes proportionately. Hence for moderate sweep rates, the capacitance is independent of the sweep rate. This figure is drawn according to Kampouris et al. (2015) and Gryglewicz et al. (2005). (b) Phase transition at a certain voltage potential. Polarization voltage causes the drift of the peaks.

9.3.3 Current pulse method

Current pulses can be used to detect the DL capacitance [Benger et al. (2009)]. For this purpose, a specific DC pulse is applied, which loads the cell or is taken from the cell. An example is presented in Fig. 9.9. In this case, a provided current loads the cell. Therefore, U_{cell} increases. From the specific form of the voltage response, a statement about the DL capacitance can be made. The internal resistance R_i is responsible for the instantaneous step in voltage (see Eq. (9.33)). Subsequently, U_{cell} increases less strongly, caused by the charge transfer and DL capacitance. The charge transfer resistance R_{ct} can be calculated using Eqs. (9.34) and (9.35). The time constant τ in Eq. (9.36) depends on the time interval δt, which has been set to a certain value. Eventually C_{dl} is the ratio of τ to R_{ct} [Benger et al. (2009); Waag et al. (2013)].

$$R_i = (U_1 - U_0)/I, \tag{9.33}$$

$$R_{\text{tot}} = (U_2 - U_0)/I, \tag{9.34}$$

$$R_{\text{ct}} = R_{\text{tot}} - R_i, \tag{9.35}$$

$$\tau = \frac{\delta t}{\ln\left(1 - \frac{U_3 - U_4}{U_3 - U_{\text{end}}}\right)}, \tag{9.36}$$

$$C_{\text{dl}} = \tau / R_{\text{ct}}. \tag{9.37}$$

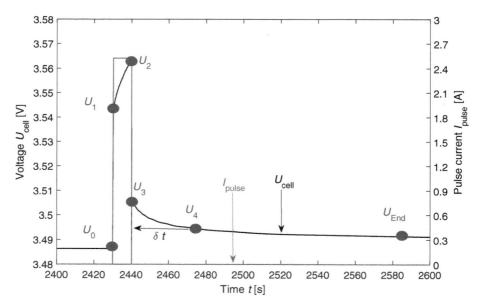

Fig. 9.9 Provided pulse current I and measured voltage response U of a Li-ion cell. Retraced according to Benger et al. (2009).

Again, the cell must be in equilibrium before measurement (a constant voltage U_0 before turning on the pulse current, as seen in Fig. 9.9). Caution should be exercised when comparing calculated values from this measurement to the results from EIS measurement, due to the different measurement techniques [Waag et al. (2013)].

9.3.4 Summary

EIS is a widely used measurement technique for the identification of frequency-based processes in cells, and it can be used to estimate the State of Health (SoH) of a cell. Bard and Faulkner (1980) transform and assign the measured impedance $Z(\omega)$ from EIS to elements from a Randles circuit. The authors present chemical concentration gradients in a RedOx system as a cause for pseudocapacitive effects [Bard and Faulkner (1980)]. A transformation of the data, measured by EIS, from the frequency domain into the time domain by means of the distribution of relaxation times (DRT) can reveal additional information about cells [Waag et al. (2013)]. The advantage of DRT is the separation of time constants of different processes [Schmidt et al. (2013)].

Regarding mobile battery systems, superposed DC currents (if the cell is under load) and the impact of the battery system itself (additional electronics, casing, cables, etc.) on impedance could superpose the excitation signal using EIS. Furthermore, many cells in a battery system could overlap their response signals. The disturbances could reduce the relevant information from the response signal or make it uninterpretable. In contrast to the EIS, the current pulse method makes online measuring easier to realize since cycle profiles of EVs, for instance, include power pulses as well [Benger et al. (2009)]. However, the resolution time of logging must be very high (in the low millisecond range) [Benger et al. (2009)] in order to measure appropriately.

9.4 Utilization of Capacitive Effects in Li-ion Cells

9.4.1 Li-ion cell development

One can speak of two main optimizations in cell production: high-energy and high-power. Particularly for mobile applications, energy density is a key attribute for the successful electrification of vehicles. In the case of electric vehicles (EVs) with large battery capacity (in Ah), the power demand is provided by this large battery capacity.[13] High cell capacities mean high

---

[13]Meaning that a single cell does not need to provide high currents since many strings of cells are connected in parallel.

achievable currents. However, some devices or products like power tools or hybrid electric vehicles (HEVs) need high peak currents and a small installation space for the battery. In this case, high-power cells with sufficient energy density are used. The power density is usually maintained, considering they have the same cell chemistry as high-energy cells, by a larger electrode-electrolyte interface through more layers of thinner electrodes and sufficiently dimensioned current collectors. Other ways to improve power density are (1) using electrodes with higher porosities, (2) a higher mass ratio of conductive additive to active material, which increases electric conductivity, and (3) other active materials with high-power capabilities like electrodes with a spinel structure (e.g., $LiMn_2O_4$ (LMO), $Li_4Ti_5O_{12}$ (LTO)). Compared to high-energy cells, the variant with thinner layers of active material and thicker current collectors, with respect to the total electrode thickness, does not alter the electrochemical function of the cell. Nevertheless, it can ensure that higher currents can be applied because (1) the ions have to diffuse through less active material and (2) the cell surface area is increased due to more electrode layers. Due to the larger surface area (electrode-electrolyte interface) of high-power cells, the DL capacitance should also increase. In extreme cases, a cell can be designed such that a hybrid of a capacitor and an electrochemical cell arises, as will be further elaborated upon in the next section.

9.4.2 Li-ion capacitor

Li-ion capacitors are asymmetric cells that include a negative intercalation electrode for intercalation of Li-ions, and a positive electrode for DL charge accumulation. The positive electrode usually consists of activated carbons with respective high surface areas. The principle of operation of the Li-ion capacitor is as follows: The negative intercalation electrode, such as LTO or graphite, stores Li^+ Faradaicly, whereas the anions accumulate at the carbon electrode. The Li^+ lowers the electrode voltage versus reference potential Li/Li^+, which leads to a higher energy density as compared to DLCs. Consequently, negative electrodes with relatively low potential of the intercalation process versus lithium reference potential (Li/Li^+) must be used, such as LTO [Amatucci et al. (2001)]. The combination of a nanostrucutred LTO electrode and activated carbon as a positive electrode from Amatucci et al. (2001) shows high current capabilities and a high cycle life.

A major difference between Li-ion capacitors and Li-ion cells is the change of ion concentration in the electrolyte, with respect to Li-ion capacitors.[14] Both the Li$^+$ and the anions are involved as active species in the charge separation in a Li-ion capacitor.

9.4.3 Estimation of DL capacitance on cell level

Table 9.1 shows the electrodes' length, width and thickness of a *Panasonic NCR18650B* cell based on data from Hagen *et al.* (2015). This cell consists of lithiated graphite Li$_x$C$_y$ (Gr) and lithium-nickel-cobalt-aliminium-oxide Li$_{1-x}$NiCoAlO$_2$ (NCA) [Hagen *et al.* (2015)]. In this calculation, it is assumed that the conductive improver carbon with a specific surface area of $S/m = 65 \, \text{m}^2\text{g}^{-1}$, obtained by the BET method, is in the cell. The *BET method* is a technique to measure electrodes' porosities.[15] Since nitrogen molecules are smaller than Li and solvated Li-ions, the BET method cannot predict the actual electrode surface moistened by electrolyte. Nonetheless,

Table 9.1 (*Upper half*): Data of electrodes of the Panasonic NCR18650B cell from [Hagen *et al.* (2015)]. (*Lower half*): Assumed BET values of specific material and assumed mass ratios of conductive additive (carbon) to electrode material (without current collector). The effective BET values of the electrodes are calculated as the sum of specific BET values (material dependent) times the mass ratios.

Characteristic	Li$_{1-x}$NiCoAlO$_2$ (NCA)	Li$_x$C$_y$ (Gr)
Thickness (single side)^a [μm]	82.5	95
Length [mm]	548	610
Width [mm]	61	61
Planar area (double coated) [cm^2]	668.56	744.2
Electrode mass incl. cur. collect. [g]	20	14.4
Electrode mass m_elec^{ab} [g]	18.6	11.1
Mass fraction $m_\text{active}/m_\text{elec}$ [%]	86	93
Mass fraction $m_\text{carbon}/m_\text{elec}$ [%]	8	5
BET carbon $S_\text{carbon}/m_\text{carbon}$^{c} [m^2/g]	65	65
BET active material $S_\text{active}/m_\text{active}$ [m^2/g]	0.5	4
Effective electrode BET $S_\text{eff}/m_\text{elec}$ [m^2/g]	5.63	6.97
Effective area S_eff^{d} [m^2]	104.7	77.4

^{a}Without current collector.
^{b}Including binder mass.
^{c}Based on Rhee *et al.* (2005) and IMERYS Graphite & Carbon (2017).
^{d}Scaled to the mass of the 18650 cell.

---

[14]The main difference: Li-ion capacitors have only one intercalation electrode compared to Li-ion cells.
[15]Measurement method: Pressure difference of volume flow before and after the sample due to the adsorption of nitrogen molecules on a porous sample.

the BET values are taken here for a rough estimation. Furthermore, we assume comprehensible mass fractions of carbon to total electrode $m_{\text{carbon}}/m_{\text{elec}}$ and of active material to total electrode $m_{\text{active}}/m_{\text{elec}}$.[16] Since metal oxide conductivity is poorer as compared to graphite,[3] the ratio of the conductive additive to the total mass of the positive electrode is assumed to be higher than the respective ratio of the negative electrode, in order to sustain sufficient electric conductivity of the cell (see Table 9.1). We set the BET-values of the active material and conductive additive per electrode in the ratios (see Eq. (9.38)) and thus estimated the effective surface enlargement of the electrodes by the additive carbon. The effective areas S_{eff} in Table 9.1 are the results of multiplying the effective BET values per electrode $S_{\text{eff}}/m_{\text{elec}}$ with the corresponding mass m_{elec} per electrode, as shown in Eq. (9.38).

$$S_{\text{eff}} = \left(\frac{S_{\text{carbon}}}{m_{\text{carbon}}} \cdot \frac{m_{\text{carbon}}}{m_{\text{elec}}} + \frac{S_{\text{active}}}{m_{\text{active}}} \cdot \frac{m_{\text{active}}}{m_{\text{elec}}} \right) \cdot m_{\text{elec}}. \qquad (9.38)$$

We simplify the BDM model from Sec. 9.1.1 (see Fig. 9.1) and neglect the diffusive ion layer in the electrolyte and pseudocapacitive effects. Since there is a SEI on lithiated graphite due to the low electrochemical potential versus Li/Li$^+$, we take into account the SEI as dielectric similar to the inner Helmholtz layer from the BDM model. The actual processes at the interfaces — graphite-SEI, SEI-electrolyte and processes within the SEI — are complex and partly not understood. In order to reduce complexity, a single species SEI with lithium carbonate is assumed. The permittivity of the reaction product lithium carbonate is assumed to be $\epsilon_{\text{SEI}} = 4.3 \cdot 10^{-11} \text{CV}^{-1}\text{m}^{-1}$ [Colclasure et al. (2011)]. Within the SEI, a voltage gradient is formed, which is influenced by charge concentrations (electron and ion interstitials, and migrated ions from the electrolyte) and diffusion processes [Colclasure et al. (2011)]. Colclasure et al. (2011) note a Debye length of $0.66 < \lambda_D < 2$ nm in case of electron concentrations $1.3 \cdot 10^{-3} < C_E < 1.3 \cdot 10^{-2}$ kmol m^{-3}. The Debye length expresses how far the electrostatic field reaches until it has weakened by the factor $1/e$. Figure 9.10 addresses the voltage drop across the SEI and the corresponding Debye length. Assuming a linear gradient now in molar concentrations of Li$^+$ and e$^-$ interstitials[17] [Colclasure et al. (2011)], the Debye length and hence the voltage drop across the SEI could

---

[16]The remaining mass fraction $1 - m_{\text{carbon}}/m_{\text{elec}} - m_{\text{active}}/m_{\text{elec}}$ is binder.

[17]A dielectric is charge neutral, meaning the same concentrations of Li$^+$ and e$^-$ [Colclasure et al. (2011)].

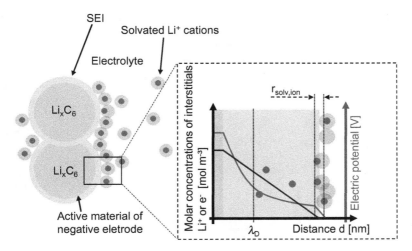

Fig. 9.10 (*Left*) Schematic drawing of SEI on lithiated graphite as particles. (*Right*) Molar concentrations of charge carriers and corresponding voltage drop depending on distance d from the electrode–SEI interface. Left particles and the molar concentration curve retraced according to Colclasure et al. (2011).

correlate with $\sqrt{1/C_E}$, referring to Eq. (9.39). It is now assumed in the following that these processes shrink the effective charge separation distance to approximately $d_{\text{SEI,eff}} = 3\,\text{nm}$, despite the thickness of the SEI between 3 to 100 nm [An et al. (2016)].

$$\lambda_D = \sqrt{\frac{\epsilon_r \epsilon_0 RT}{2F^2 C_E}}. \tag{9.39}$$

The inner specific Helmholtz capacitance C_{in}/S across the SEI is calculated using Eq. (9.40). In addition, by considering the radii of solvated ions as the outer Helmholtz layer, the outer specific capacitance per area C_{out}/S is estimated here based on the assumed radii $r_{\text{solv.ion}}$ of solvated ions (here Li$^+$ and hexafluorophospate PF$_6^-$) and the permittivity of electrolyte ϵ_{mix} using Eq. (9.41) [Zhang et al. (2009)].

$$C_{\text{in}}/S = \epsilon_{\text{SEI}}/d_{\text{SEI,eff}}, \tag{9.40}$$

$$C_{\text{out}}/S = \epsilon_0 \epsilon_{\text{mix}} \cdot 1/r_{\text{solv.ion}}. \tag{9.41}$$

We consider the common solvents ethylene carbonate (EC) and diethyl carbonate (DEC) in the mass ratio 1:1. The individual permittivities of the solvents are taken from Xu (2004). The volume ratio X_{DEC} is the quotient of the DEC volume to total mixture volume (without conductive salt). Then,

Table 9.2 Estimation of DL capacitance by means of Eq. (9.43). We neglect possible film layers on the positive electrode. The distance of the charge separation at the positive electrode is assumed to be equal to the radius of solvated PF_6^-.

Value	$Li_{1-x}NiCoAlO_2$ (NCA)	Li_xC_y (Gr)
$r_{\text{solv.ion}}$ PF_6^- | Li^+ [a] [m]	$2 \cdot 10^{-9}$	$0.4 \cdot 10^{-9}$
Permittivity ϵ_{EC} [b] []	89.78	89.78
Permittivity ϵ_{DEC} [b] []	2.805	2.805
Permittivity ϵ_{mix} EC:DEC (1:1) []	40	40
C_{out}/S [$\mu F/cm^2$]	17.7	88.5
Thickness of dielectric (SEI) [m]	0	$3 \cdot 10^{-9}$
C_{in}/S [$\mu F/cm^2$]	–	1.43
C_{dl}/S [$\mu F/cm^2$]	17.7	1.41
$C_{\text{dl,Seff}}$ [F]	18.53	1.09
$C_{\text{dl,18650}}$ [F]		1.03

[a] Values based on Parsons (2014).
[b] At 25°C, from Xu (2004).

the permittivity of the mixture is estimated by means of Eq. (9.42), according to [Xu (2004)].

$$\epsilon_{\text{mix}} = (1 - X_{\text{DEC}})\epsilon_{EC} + X_{\text{DEC}}\epsilon_{DEC}. \quad (9.42)$$

Taking these assumptions into account, the voltage drop across DL is qualitatively presented in Fig. 9.10 as the sum of inner and outer Helmholtz layers. Hence, the total DL capacitance of the negative electrode is described here as a serial connection of two capacitances.[18]

$$C_{\text{dl}} = \frac{C_{\text{in}} \cdot C_{\text{out}}}{C_{\text{in}} + C_{\text{out}}}. \quad (9.43)$$

We see as a result in Table 9.2 a significantly higher DL capacitance for the positive electrode, since we disregard any surface layers on that electrode. The specific capacitances C_{dl}/S of Gr and NCA are estimated as $1.41\,\mu F\,cm^{-2}$ and $17.7\,\mu F\,cm^{-2}$. The specific capacitance of graphite on a smooth surface is around $C_{\text{dl}}/S = 20\,\mu F\,cm^{-2}$ [Shi (1996)]. The specific capacitance of DLCs based on carbon materials is around $C_{\text{dl}}/S = 10 - 20\,\mu F\,cm^{-2}$ [Zhang et al. (2009)]. The cause of the different values between our calculated C_{dl}/S for Gr and those in the literature could be,

---

[18] These two capacitances would have to be modeled as two RC-circuits with different time constants, as EIS measurements revealed. In this calculation, the time constants are assumed to be the same.

e.g., the disregard of the SEI (no lithiated graphite), other permittivities or a different measurement method to measure surface areas (micropores, etc.) of the specific material. The specific capacitance per electrode C_{dl}/S is then multiplied by S_{eff} from Table 9.1 to obtain the capacitance per electrode $C_{dl,Seff}$ in Table 9.2. The total capacitance of the cell with $C_{dl,18650} = 1.03\,\text{F}$ is calculated analogously to Eq. (9.43), using the two capacitance values of the electrodes $C_{dl,Seff}$ instead of C_{in} and C_{out} [Choi et al. (2012)].[19]

There is probably even more potential for DL capacitance on the cell level, in the case of cells with larger interface area (assuming the same electrode materials). High-power cells often have more layers, or windings, than energy-optimized cells. A twice as large electrode surface area seems realistic, probably resulting in a doubled DL capacitance $C_{dl,HP,18650} = 2.06\,\text{F}$ based on the results from Table 9.2. However, the thinner electrode thickness reduces the capacity of high-power Li-ion cells, which is estimated here by a corresponding 50% reduction of Q_n.

Another way to increase cell surface area is changing the cell format from a cylindrical to a prismatic one. The DL capacitance of a 25 Ah prismatic high-power Li-ion cell is estimated here over the ratio of cell capacities (25 Ah to 1.6 Ah).[20]

$$C_{dl,HP,prism} = \frac{Q_{n,HP,prism}}{Q_{n,HP,18650}} \cdot C_{dl,HP,18650} = \frac{25}{1.6} \cdot 2.06\,F = 32.19\,F.$$
(9.44)

9.4.4 Potential on the system level

The Li-ion cells must be connected in series to a *string* in order to achieve the appropriate system voltage. The number of strings in parallel sets the capacity [Ah] of the battery system. Hence, the cell topology could limit or enhance cell capacitance effects [F], as the equations of total capacitance in a series and parallel circuit reveal. Table 9.3 shows scaled DL capacitance values only by the number of serial- and parallel-connected cells, based on the calculation on the cell level from Sec. 9.4.3. The system requirements *System voltage* and *Energy* are set to a typical value based on the application fields EV, HEV and *Stationary energy storage* (STAT). The capacity [Ah] and thus the necessary number of strings and number of series-connected cells depend

---

[19]The intention of Choi et al. (2012) is to calculate the total capacitance of a DLC.
[20]Prismatic cell: 50 Ah high-energy | 25 Ah high-power; Panasonic 18650 cell from Table 9.1: 3.2 Ah high-energy | 1.6 Ah high-power.

Table 9.3 Scaled DL capacitance values [F] for three application fields based on results from Table 9.2 and Eq. (9.44). The results refer to two cell types: prismatic and 18650 of the type high-power or high-energy. The considered cell characteristics: High-energy 18650: $Q_{n,18650} = 3.2\,\text{Ah}$; $C_{dl,18650} = 1.03\,\text{F}$. High-power 18650 for HEV: $Q_{n,HP,18650} = 1.6\,\text{Ah}$; $C_{dl,HP,18650} = 2.06\,\text{F}$. High-energy prismatic: $Q_{n,prism} = 50\,\text{Ah}$; $C_{dl,prism} = 16.09\,\text{F}$. High-power prismatic for HEV: $Q_{n,HP,prism} = 25\,\text{Ah}$; $C_{dl,HP,prism} = 32.19\,\text{F}$. All cells have the same chemistry, meaning the same nominal voltage $U_{n,18650} = U_{n,prism} = 3.6\,\text{V}$.

System	System voltage [V]	Energy [kWh]	Capacity [Ah]	$C_{dl,18650}$ [F]	$C_{dl,prism}$ [F]
EV	400	90	225	0.65	0.72
HEV	48	2.4	50	4.71	4.60
STAT	400	1000	2500	7.19	7.18

on the considered cell type. The use cases EV and STAT have high-energy cells whereas high-power cells are being considered for HEV.[21]

In contrast to EVs, there is a larger DL capacitance [F] suspected in HEVs caused by a lower system voltage, as seen in Table 9.3. A further result of this calculation is that the influence of cell type (prismatic versus cylindrical) is reduced on the system level, as compared to the cell level.

Application field 48 V

High frequent, steep currents can arise in mobile applications like HEV through boosting/recuperation mode or mobile communication. Now the idea is to find a possible application field where the considered conventional Li-ion batteries supply the energy only by their superposed DL capacitance, without stressing the cells electrochemically. Car tracking with GPS and GSM systems with sensor data inside the vehicle requires robust and powerful batteries [VARTA Microbattery GmbH (2016)]. Telecommunication profiles similar to the profile in Fig. 9.11 could be exemplarily supplied by the superposed DL capacitance of Li-ion cells [Ong and Newman (1999)].

We focus on a HEV system using 18650 cells and 48 V as a nominal voltage. We have calculated $C_{dl,HEV} = 4.71\,\text{F}$ based on 18650 high-power cells from Table 9.3, in which we have assumed 14 identical cells in series. Electrotechnically, the DL capacitance acts in parallel with the charge

---

[21] A possible business case is the second-life concept of aged high-energy cells from EVs. As a result of aging, the cells experience a loss of capacity. However, EVs require a sufficiently high-energy density. If the cells no longer meet the requirements, they can be re-used in a STAT.

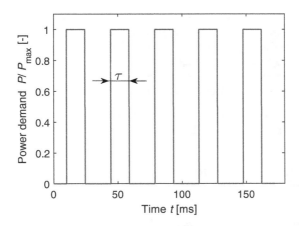

Fig. 9.11 Exemplary GSM communication profile.

transfer resistance interpreting a Randles circuit (see Sec. 9.3.1.1). Assuming a charge transfer resistance of a 18650 cell with $R_{ct} = 7\,\text{m}\Omega$, the resistance of the system could be $R_{ct,\text{HEV}} = 3.06\,\text{m}\Omega$ (32 strings in parallel, each string has 14 cells) for homogeneous cells with identical R_{ct} and C_{dl} per cell. The time constant of the superposed DL capacitance on the system level would then be:

$$\tau = C_{dl,\text{HEV}} \cdot R_{ct,\text{HEV}} = 4.71\,\text{F} \cdot 0.00306\,\Omega = 14.4\,\text{ms}. \tag{9.45}$$

Assuming the power consumption of the data transfer with GPRS, according to Paller *et al.* (2015), with approximately 100 mA as the peak current consumption, the assumed release of charge by the DL is given by Eq. (9.1) with maximum potential difference of the RC-circuit[22] $\Delta U_{\text{RC,HEV}}$:

$$\Delta Q_{\text{HEV}} = \Delta U_{\text{RC,HEV}} \cdot C_{dl,\text{HEV}}, \tag{9.46}$$

$$= \Delta I_{ct,\text{HEV}} \cdot R_{ct,\text{HEV}} \cdot C_{dl,\text{HEV}}, \tag{9.47}$$

$$= 0.1\,\text{A} \cdot 0.00306\,\Omega \cdot 4.71\,\text{F} = 1.44\,\text{mAs}. \tag{9.48}$$

A typical end of life criterion for mobile applications is the increase of internal resistance by a factor of 2. As R_{ct} increases with cell aging, the DL capacitance can probably also take up more charge in already-aged cells, for example.

---

[22]RC-circuit: A capacitance and a resistance connected in parallel.

Table 9.4 Storable amount of charge ΔQ_{HEV} by DL capacitance for the exemplary HEV topology. Variation of current $\Delta I_{ct,HEV}$ and charge transfer resistance $R_{ct,HEV}$ (R_{ct} values are doubled within the brackets to account for an aged system).

$R_{ct,HEV}$ [$m\Omega$]	Current pulse $\Delta I_{ct,HEV}$ [A]	Accumulated charge ΔQ_{HEV}
3.06 (6.12)	0.1	1.44 (2.88) mAs
3.06 (6.12)	1	14.41 (28.83) mAs
3.06 (6.12)	10	144.13 (288.25) mAs
3.06 (6.12)	100	0.40 (0.80) mAh

A GSM communication[23] needs about 2370 mAs where the information is sent via GSM with a packet size of 287 bytes [Paller et al. (2015)].[24] The comparison of our results with this amount of charge necessary for GSM communication shows that the utilization of the integrated DL capacitance in the Li-ion cell, as energy storage, is limited on the system level. Increasing the pulse current as seen in Table 9.4 could lead to a higher charge released by the DL in the Li-ion cell. However, the profile would have to be adapted exactly to the internal capacitance. The power demand of communication needs to be adapted in shape and frequency in order to activate the DL. Neither pseudocapacitive effects like surface-near intercalation nor adsorption are regarded. The potential of capacitive effects could be possibly significantly increased if pseudocapacitive effects are taken into account. According to Conway et al. (1997), the pseudocapacitance can be up to 10–100 times the DL capacitance, focusing on suitable materials which allow high pseudocapacitance.

Consequently, the amount of DL in Li-ion systems depends on (1) system topology (number of serially- and parallel-connected cells), (2) shape of the load profile (applied current and frequency), and (3) cell characteristics such as R_{ct}, cell format (capacity per cell), cell development (high-energy, high-power), electrolyte and cell material (type of carbon and active material).

Cell technologies for 48 V applications

Table 9.5 lists the already introduced storage technologies: The *Li-ion cell*, *DLC* and *Li-capacitor* by their capacitance. The table is based on 48 V,

---

[23] A system with low energy consumption according to Paller et al. (2015).
[24] A size of 100 bytes per message is sufficient for cloud-based battery management control [Khayyam et al. (2013)].

Table 9.5 Comparison of typical values on 48 V between a DLC, a Li-capacitor and a high-power Li-ion battery from Table 9.3. Energy values are based on the voltage range from 42 V to 58 V. NxM stands for the number of cells in series (letter N) and number of strings in parallel (letter M).

System	Cells (NxM) [-]	Capacitance [F]	Capacity [Ah]	Energy [Wh]	Energy density [Wh kg^{-1}]
Li-ion HP 18650	14x32	4.71	50	2400	114[a]
Li-capacitor	16x1	93[b]	0.41[c]	20.7	4[b]
DLC	18x1	12.8[b]	0.06[c]	2.8	0.5[b]

[a] Assumption: 448 *Panasonic NCR18650B* cells with 47.1 g per cell.
[b] Data based on Eger Porsche AG (2017).
[c] $\Delta Q = \Delta U \cdot C = (58\,\text{V} - 42\,\text{V}) \cdot C = 16\,\text{V} \cdot C$

a typical voltage class for partially electrified drive systems. We set the number of serially-connected Li-ion cells again to 14 for the exemplary HEV topology. For a cell voltage range of $3\,\text{V} \leq U_\text{cell} \leq 4.15\,\text{V}$, the voltage on 48 V varies between 42 V to 58 V. The energies of the other two technologies are scaled to the same voltage range. A Li-capacitor can achieve $C_\text{dl} = 93\,\text{F}$ [Eger Porsche AG (2017)] with a charge accumulation of $\Delta Q = 400\,\text{mAh}$. Despite the high capacitance of the Li-capacitor, only a small amount of energy can be stored as compared to conventional Li-ion cells. Interpreting the DL electrotechnically with the Randles circuit, a sufficiently high resistance in parallel to C_dl is needed to activate the DL capacitance of a Li-ion cell [Ong and Newman (1999)]. The energy of the superposed capacitance in a Li-ion cell, as well as in the system (see Eqs. (9.50) and (9.51)), could correlate to the storable energy of a usual capacitor (see Eq. (9.49)), which is addressed in the following equations.

$$E = 1/2 \cdot C \cdot (U_\text{up}^2 - U_\text{lo}^2), \tag{9.49}$$

$$E_\text{dl} = 1/2 \cdot C_\text{dl} \cdot (U_0^2 - (U_0 - \Delta U_\text{RC})^2), \tag{9.50}$$

$$E_\text{dl,HEV} = 1/2 \cdot C_\text{dl,HEV} \cdot (U_{0,\text{HEV}}^2 - (U_{0,\text{HEV}} - \Delta U_\text{RC,HEV})^2). \tag{9.51}$$

The variable U_0 is the relaxed cell voltage before the pulse current and ΔU_RC, the voltage drop accounting for the DL capacitance in the cell. With following relations (assumptions) in Eqs. (9.52) to (9.55), Eq. (9.56) is derived from Eq. (9.51), where N and M are the numbers of connected cells in series and strings in parallel.

$$C_\text{dl,HEV} = MC_\text{dl}/N, \tag{9.52}$$

$$U_{0,\text{HEV}} = U_0 N, \tag{9.53}$$

$$\Delta I_{\text{ct,HEV}} = \Delta I_{\text{ct}} M, \tag{9.54}$$

$$R_{\text{ct,HEV}} = R_{\text{ct}} N/M, \tag{9.55}$$

$$E_{\text{dl,HEV}} = MN C_{\text{dl}} \Delta I_{\text{ct}} R_{\text{ct}} (U_0 - \Delta I_{\text{ct}} R_{\text{ct}}/2), \tag{9.56}$$

$$G_{\text{HEV}} = MN Q_{\text{n}} U_{\text{n}}. \tag{9.57}$$

Because the cells must be connected in series with N cells in order to achieve a desired system voltage, and with M strings already in parallel in order to achieve a desired system capacity, the energy stored by the integrated DL on the system level is expected to increase linearly with N and M, referring to Eq. (9.56). Equation (9.57) shows the Faradaic energy G of the Li-ion cells on the system level, with nominal cell capacity Q_{n} and nominal cell voltage U_{n} as comparison. Taking note of Eqs. (9.56) and (9.57), the ratio of the stored energy (only by DL capacitance) to the totally storable energy $E_{\text{dl,HEV}}/G_{\text{HEV}}$ stays constant with an increasing system size.

9.5 Conclusion and Outlook

DLCs and Li-ion cells have many similarities. Polarized boundary layers form a superposed DL in Li-ion cells. These DL capacitances can occur in parallel to pseudocapacitive effects on a time scale. In contrast to DL capacitance, pseudocapacitive effects can be caused by mass transfer on surface-near intercalation sites in the electrodes. The rough calculation of the superposed DL capacitance in Li-ion cells results in values around 2 F for a 1.6 Ah high-power 18650 cell (and even around 32 F for a prismatic 25 Ah high-power cell, neglecting surface-near intercalation and adsorption). The estimation of the DL capacitance on a characteristic HEV system level (48 V, 50 Ah, 2400 Wh) yields 4.7 F. However, the amount of charge only stored by the DL capacitance is limited and has to be activated through adapted current pulses. Next, the ratio of energy stored by the superposed DL to the total energy (capacity times voltage) seems to stay constant with increasing storage size (increasing energy). The estimated values (capacitances, amounts of charge released by DL and cell specific characteristics) are partly based on rough calculations and need to be validated by experimental work. A promising measurement method for this purpose of identification and quantification of capacitive effects in Li-ion cells is EIS.

Cell designers today are already using thinner electrodes with larger cell surface areas to obtain lower internal cell resistance and to make cells more applicable to high-power applications. With respect to high-power cell development, the potential of capacitance effects in Li-ion cells could increase. An open question is to whether cell degradation can be reduced by the utilization of capacitance effects (surface-near intercalation and DL capacitance), considering special application fields where short-term currents are required. If this is possible, the question arises as to what extent cell aging can be minimized through optimized cell design.

Nomenclature

F	Faraday constant: 96485.33289 Asmol^{-1}
k_B	Boltzmann constant: $1.38064852 \cdot 10^{-23}$ JK^{-1}
R	Gas constant: 8.3144598 kg m^{2}s^{-2}K^{-1}mol^{-1}
$\Delta I_{ct,HEV}$	Maximum (pulse) current across R_{ct} on 48 V [A]
ΔI_{ct}	Maximum (pulse) current across R_{ct} on cell level [A]
ΔQ	Separated and accumulated charges [Ah]
ΔQ_{HEV}	Estimated charge accumulation through DL in HEV [Ah]
δt	Time interval [s]
ΔU	Potential difference [V]
$\Delta U_{RC,HEV}$	Maximum potential difference of RC-circuit on 48 V [V]
ΔU_{RC}	Maximum potential difference of RC-circuit on cell level [V]
δ	Density [kg m^{-3}]
$\Delta \phi$	Phase angle between current and voltage [°]
ϵ_{mix}	Permittivity of mix of solvents [-]
ϵ_r	Relative permittivity of a dielectric medium [-]
ϵ_{SEI}	Dielectric of SEI [Fm^{-1}]
$\hat{i}$	Current amplitude [A]
$\hat{u}$	Voltage amplitude [V]
κ^{-1}	Thickness of outer Helmholtz layer [Å]
λ_D	Debye length [m]
ω	Radian frequency [Hz]
ϕ_i	Phase angle of current i [°]
ϕ_u	Phase angle of voltage u [°]
ϕ_w	Phase angle of Warburg impedance [°]
ϕ_{ne}	Potential of negative electrode versus reference [V]
ϕ_{pe}	Potential of positive electrode versus reference [V]
σ	Mass transfer coefficient [ω s$^{-1/2}$]
σ_{con}	Electrical conductivity [S m^{-1}]

τ	Time constant [s]
Θ	Occupied lattice sites [-]
Θ_s	A free lattice site [-]
ϵ_0	Vacuum permittivity: $8.854188 \cdot 10^{-12}\,\text{F m}^{-1}$
A	Planar electrode area [cm^2]
C	Capacitance [F]
C'	Real part of capacitance [F]
C''	Imaginary part of capacitance [F]
C/S	Specific capacitance per surface area [F m^{-2}]
C_{dl}	Superposed double-layer capacitance in Li-ion cell [F]
C_{dl}/m	Specific double-layer capacitance [F g^{-1}]
C_{init}	Capacitance for charge accumulation [F]
C_O	Concentration of species Ox [mol m^{-3}]
C_{plate}	Capacitance plate capacitor [F]
C_R	Concentration of species Red [mol m^{-3}]
C_s	Pseudocapacitance [F] according to *Bard* and *Bruce et al.*
C_w	Capacitance of Warburg impedance [F]
C_O^*	Bulk concentration of species Ox [mol m^{-3}]
C_R^*	Bulk concentration of species Red [mol m^{-3}]
$C_{\text{dl},18650}$	Estimated DL capacitance of 18650 cell [F]
$C_{\text{dl,HEV}}$	Estimated DL capacitance of hybrid electric vehicle [F]
$C_{\text{dl,HP},18650}$	Estimated DL cap. of 1.6 Ah 18650 high-power cell [F]
$C_{\text{dl,HP,prism}}$	Estimated DL cap. of 25 Ah prismatic high-power cell [F]
$C_{\text{dl,prism}}$	Estimated DL capacitance of prismatic cell [F]
C_E	Molar concentration of Li$^+$ or e$^-$ interstitials [mol m^{-3}]
C_{in}	DL capacitance of inner Helmholtz layer [μF cm^{-2}]
C_{out}	DL capacitance of outer Helmholtz layer [μF cm^{-2}]
C_ϕ	Pseudocapacitance [F] according to *Conway*
d	Distance [m]
D_O	Diffusion coefficient of species Ox [m^2 s^{-1}]
D_R	Diffusion coefficient of species Red [m^2 s^{-1}]
$d_{\text{SEI,eff}}$	Effective distance of charge separation by the SEI [m]
E	Energy [VAs]
E/m	Energy density [Wh kg^{-1}]
$E_{\text{dl,HEV}}$	Energy stored in DL capacitance in HEV [VAs]
E_{dl}	Energy stored by DL capacitance on cell level [VAs]
f	Frequency [Hz]
G	Gibbs energy [J]
G_{HEV}	Faradaic energy stored in HEV [VAs]

i	Alternating current [A]
i_0	Exchange current [A]
i_f	Faradaic current [A]
j	Imaginary unit [-]
I_{pulse}	Pulse current [A]
M	Number of strings in parallel [-]
m_{elec}	Electrode mass without current collector [g]
N	Number of serially connected cells [-]
N_i	Concentration of an ion i [mol m^{-3}]
p	Particle size [m]
Q	Charge [Ah]
Q_n	Nominal cell capacity [Ah]
R_{ad}	Adsorption resistance [Ω]
R_{ct}	Charge transfer resistance of Li-ion cell [Ω]
R_i	Internal cell resistance [Ω]
R_{lat}	Intercalation resistance [Ω]
$r_{\text{solv,ion}}$	Radius of solvated ion [m]
R_s	Series resistance [Ω]
R_{tot}	Total cell resistance [Ω]
R_w	Real part of Warburg impedance [Ω]
$R_{\text{ct,HEV}}$	Estimated charge transfer resistance in HEV [Ω]
S	Electrode interface area [cm^2]
S/m	Specific surface area [m^2 g^{-1}]
S_{eff}	Surface area scaled to planar area of a 18650 cell [m^2]
sr	Sweep rate [V s^{-1}]
T	Temperature [K]
t	Time [h]
u	Alternating voltage [V]
$U_{0,\text{HEV}}$	Relaxed cell voltage on 48 V [V]
U_0	Relaxed cell voltage [V]
U_{cell}	Cell voltage [V]
U_{lo}	Lower voltage limit [V]
U_{up}	Upper voltage limit [V]
X_{DEC}	Volume ratio of DEC to total volume [-]
X_w	Imaginary part of Warburg impedance [Ω]
z	Charge number [-]
Z'	Real part of impedance Z [Ω]
Z''	Imaginary part of impedance Z [Ω]
$\underline{Z}(\omega)$	Frequency-dependent impedance [Ω]

$\underline{Z}_\mathrm{CPE}$ Constant phase element [Ω]
$\underline{Z}_\mathrm{f}$ Frequency dependent elements [Ω]
z_i Charge number of an ion [-]
$\underline{Z}_\mathrm{w}$ Warburg impedance [Ω]

Bibliography

Amatucci, G. G., Badway, F., Du Pasquier, A., and Zheng, T. (2001). An asymmetric hybrid nonaqueous energy storage cell, *J. Electrochem. Soc.* **148**, 8, pp. A930–A939, doi:10.1149/1.1383553, http://jes.ecsdl.org/content/148/8/A930.short.

An, S. J., Li, J., Daniel, C., Mohanty, D., Nagpure, S., and Wood, D. L. (2016). The state of understanding of the lithium-ion-battery graphite solid electrolyte interphase (SEI) and its relationship to formation cycling, *Carbon* **105**, Supplement C, pp. 52–76, doi:https://doi.org/10.1016/j.carbon.2016.04.008, http://www.sciencedirect.com/science/article/pii/S0008622316302676.

Bard, A. and Faulkner, L. (1980). *Electrochemical Methods: Fundamentals and Applications* (Wiley), ISBN 9780471055426, https://books.google.de/books?id=Rfso AAAAYAAJ, pp. 27, 322–330, 350–352.

Benger, R., Wenzl, H., Beck, H. P., Jiang, M., Ohms, D., and Schaedlich, G. (2009). Electrochemical and thermal modeling of lithium-ion cells for use in HEV or EV application, *World Electr. Veh. J.* **3**, 1, pp. 342–351.

Bruce, P. and Saidi, M. (1992). *J. Electroanal. Chem.* **322**, 1, pp. 93–105, doi:http://dx.doi.org/10.1016/0022-0728(92)80069-G, http://www.sciencedirect.com/science/article/pii/002207289280069G.

Choi, N.-S., Chen, Z., Freunberger, S. A., Ji, X., Sun, Y.-K., Amine, K., Yushin, G., Nazar, L. F., Cho, J., and Bruce, P. G. (2012). Challenges facing lithium batteries and electrical double-layer capacitors, *Angewandte Chemie International Edition* **51**, 40, pp. 9994–10024, doi:10.1002/anie.201201429, http://dx.doi.org/10.1002/anie.201201429.

Choi, Y.-M. and Pyun, S.-I. (1997). Effects of intercalation-induced stress on lithium transport through porous $LiCoO_2$ electrode, *Solid State Ionics* **99**, 3, pp. 173–183, https://doi.org/10.1016/S0167-2738(97)00253-1, http://www.sciencedirect.com/science/article/pii/S0167273897002531.

Colclasure, A. M., Smith, K. A., and Kee, R. J. (2011). Modeling detailed chemistry and transport for solid-electrolyte-interface (SEI) films in Li-ion batteries, *Electrochim. Acta* **58**, Supplement C, pp. 33–43, doi:https://doi.org/10.1016/j.electacta.2011.08.067, http://www.sciencedirect.com/science/article/pii/S0013468611013120.

Conway, B., Birss, V., and Wojtowicz, J. (1997). The role and utilization of pseudocapacitance for energy storage by supercapacitors, *J. Power Sources* **66**, 1, pp. 1–14, doi: http://dx.doi.org/10.1016/S0378-7753(96)02474-3, http://www.sciencedirect.com/science/article/pii/S0378775396024743.

Conway, B. E. (1991). Transition from supercapacitor to battery behavior in electrochemical energy storage, *J. Electrochem. Soc.* **138**, 6, pp. 1539–1548, doi:10.1149/1.2085829, http://jes.ecsdl.org/content/138/6/1539.abstract.

Daniel, C. and Besenhard, J. O. (2012). *Handbook of Battery Materials* (John Wiley & Sons), pp. 269, 272, 436.

Eger *Porsche AG* (ed.) (2017). *Future Powercaps*, Vol. Seminar SPA 342 (Stuttgarter Produktionsakademie, Stuttgart, Germany).

Gamry Instruments (2017). How cabling and signal amplitudes affect EIS results, https://www.gamry.com/application-notes/EIS/accurate-eis/. Date accessed: September 2017.

Gryglewicz, G., Machnikowski, J., Lorenc-Grabowska, E., Lota, G., and Frackowiak, E. (2005). Effect of pore size distribution of coal-based activated carbons on double-layer capacitance, *Electrochim. Acta* **50**, 5, pp. 1197–1206, http://dx.doi.org/10.1016/j.electacta.2004.07.045, http://www.sciencedirect.com/science/article/pii/S0013468604008886.

Hagen, M., Hanselmann, D., Ahlbrecht, K., Maça, R., Gerber, D., and Tübke, J. (2015). Lithium–sulfur cells: The gap between the state of the art and the requirements for high-energy battery cells, *Adv. Energy Mater.* **5**, 16.

IMERYS Graphite & Carbon (2017). Carbon blacks. c-nergy, http://www.imerys-graphite-and-carbon.com/wordpress/wp-app/uploads/2014/04/IMERYS_ME_high-purity.pdf. Date accessed: October 2017.

Kampouris, D. K., Ji, X., Randviir, E. P., and Banks, C. E. (2015). A new approach for the improved interpretation of capacitance measurements for materials utilised in energy storage, *RSC Adv.* **5**, pp. 12782–12791, doi:10.1039/C4RA17132B, http://dx.doi.org/10.1039/C4RA17132B.

Khayyam, H., Abawajy, J., Javadi, B., Goscinski, A., Stojcevski, A., and Bab-Hadiashar, A. (2013). Intelligent battery energy management and control for vehicle-to-grid via cloud computing network, *Appl. Energy* **111**, Supplement C, pp. 971–981, doi:https://doi.org/10.1016/j.apenergy.2013.06.021, http://www.sciencedirect.com/science/article/pii/S030626191300528X.

Levi, M. D. and Aurbach, D. (1997). Simultaneous measurements and modeling of the electrochemical impedance and the cyclic voltammetric characteristics of graphite electrodes doped with lithium, *J. Phys. Chem. B* **101**, 23, pp. 4630–4640, doi:10.1021/jp9701909, http://dx.doi.org/10.1021/jp9701909,http://dx.doi.org/10.1021/jp9701909.

Levi, M. D., Salitra, G., Markovsky, B., Teller, H., Aurbach, D., Heider, U., and Heider, L. (1999). Solid-state electrochemical kinetics of Li-ion intercalation into $Li_{1-x}CoO_2$: Simultaneous application of electroanalytical techniques SSCV, PITT, and EIS, *J. Electrochem. Soc.* **146**, 4, pp. 1279–1289, doi:10.1149/1.1391759, http://jes.ecsdl.org/content/146/4/1279.full.pdf+html, http://jes.ecsdl.org/content/146/4/1279.abstract.

Ong, I. J. and Newman, J. (1999). Double-layer capacitance in a dual lithium-ion insertion cell, *J. Electrochem. Soc.* **146**, 12, pp. 4360–4365, doi:10.1149/1.1392643, http://jes.ecsdl.org/content/146/12/4360.abstract.

Paller, G., Szármes, P., and Élő, G. (2015). Power consumption considerations of GSM-connected sensors in the agrodat.hu sensor network, *Sensors & Transducers* **189**, 6, pp. 52–60, http://www.sensorsportal.com/HTML/DIGEST/P_2671.htm.

Parsons, D. F. (2014). Predicting ion specific capacitances of supercapacitors due to quantum ionic interactions, *J. Colloid. Interf. Sci.* **427**, Supplement C, pp. 67–72, doi:https://doi.org/10.1016/j.jcis.2014.01.018, http://www.sciencedirect.com/science/article/pii/S0021979714000344.

Plett, G. L. (2015). *Battery Management Systems Volume I. Battery Modelling* (Artech House), ISBN 9781630810238, http://us.artechhouse.com/Battery-Management-Systems-Volume-1-Battery-Modeling-P1752.aspx, p. 304.

Qu, D. and Shi, H. (1998). Studies of activated carbons used in double-layer capacitors, *J. Power Sources* **74**, 1, pp. 99–107, doi:https://doi.org/10.1016/S0378-7753(98)00038-X, http://www.sciencedirect.com/science/article/pii/S037877539800038X.

Randles, J. E. B. (1947). Kinetics of rapid electrode reactions, *Discuss. Faraday Soc.* **1**, pp. 11–19, doi:10.1039/DF9470100011, http://dx.doi.org/10.1039/DF9470100011.

Rhee, H.-K., Nam, I.-S., and Park, J. M. (2005). *New Developments and Application in Chemical Reaction Engineering* (Elsevier), p. 424.

Schmidt, J. P., Berg, P., Schnleber, M., Weber, A., and Ivers-Tiffe, E. (2013). The distribution of relaxation times as basis for generalized time-domain models for Li-ion batteries, *J. Power Sources* **221**, Supplement C, pp. 70–77, doi:https://doi.org/10.1016/j.jpowsour.2012.07.100, http://www.sciencedirect.com/science/article/pii/S0378775312012359.

Shi, H. (1996). Activated carbons and double-layer capacitance, *Electrochim. Acta* **41**, 10, pp. 1633–1639.

Sluyters-Rehbach, M. and Sluyters, J. (1970). On the impedance of galvanic cells, *J. Electroanal. Chem. Interf. Electrochem.* **26**, 2, pp. 237–257, doi:http://dx.doi.org/10.1016/S0022-0728(70)80308-4, http://www.sciencedirect.com/science/article/pii/S0022072870803084.

Taberna, P. L., Simon, P., and Fauvarque, J. F. (2003). Electrochemical characteristics and impedance spectroscopy studies of carbon-carbon supercapacitors, *J. Electrochem. Soc.* **150**, 3, pp. A292–A300, doi:10.1149/1.1543948, http://jes.ecsdl.org/content/150/3/A292.abstract.

VARTA Microbattery GmbH (2016). Vehicle tracking. Application note, http://products.varta-microbattery.com//applications/mb_data/documents/application_note/Application_Note_Car_Tracking_en.pdf.

Vetter, J., Novk, P., Wagner, M., Veit, C., Mller, K.-C., Besenhard, J., Winter, M., Wohlfahrt-Mehrens, M., Vogler, C., and Hammouche, A. (2005). Aging mechanisms in lithium-ion batteries, *J. Power Sources* **147**, 1, pp. 269–281, doi:http://dx.doi.org/10.1016/j.jpowsour.2005.01.006, http://www.sciencedirect.com/science/article/pii/S0378775305000832.

Waag, W., Kbitz, S., and Sauer, D. U. (2013). Experimental investigation of the lithium-ion battery impedance characteristic at various conditions and aging states and its influence on the application, *Appl. Energy* **102**, pp. 885–897, http://dx.doi.org/10.1016/j.apenergy.2012.09.030, http://www.sciencedirect.com/science/article/pii/S030626191200671X.

Xu, K. (2004). Nonaqueous liquid electrolytes for lithium-based rechargeable batteries, *Chem. Rev.* **104**, 10, pp. 4303–4418, doi:10.1021/cr030203g, http://dx.doi.org/10.1021/cr030203g, pMID: 15669157.

Zaghib, K., Nadeau, G., and Kinoshita, K. (2000). Effect of graphite particle size on irreversible capacity loss, *J. Electrochem. Soc.* **147**, 6, pp. 2110–2115, doi:10.1149/1.1393493, http://jes.ecsdl.org/content/147/6/2110.full.pdf+html, http://jes.ecsdl.org/content/147/6/2110.abstract.

Zhang, Y., Feng, H., Wu, X., Wang, L., Zhang, A., Xia, T., Dong, H., Li, X., and Zhang, L. (2009). Progress of electrochemical capacitor electrode materials: A review, *Int. J. Hydrogen Energ.* **34**, 11, pp. 4889–4899, doi:http://dx.doi.org/10.1016/j.ijhydene.2009.04.005, http://www.sciencedirect.com/science/article/pii/S0360319909004984.

Chapter 10

Battery Recycling: Focus on Li-ion Batteries

Daniel Horn, Jörg Zimmermann, Andrea Gassmann,
Rudolf Stauber and Oliver Gutfleisch

10.1 Battery Materials and their Supply

In the world of batteries we distinguish between non-rechargeable (or primary batteries) and rechargeable (or secondary batteries):

Primary batteries include Zinc Carbon (ZnC), Alkaline Manganese (AlMn), Zinc Air (ZnAir), Silver Oxide (AgO) and Lithium Manganese Dioxide ($LiMnO_2$) batteries. Commercial secondary batteries include Nickel Cadmium (NiCd), Lead-Acid, Nickel Metal Hydride (NiMH), lithium-ion (LiB) and Li-ion-polymer (Li-Po) batteries.

According to Frost & Sullivan (2010), in 2009, the majority of the global revenue related to the lithium battery market was about 40%, in particular, for Li-ion batteries that accounted for 37.9%. The high demand for mobile and portable appliances makes them the most popular battery type today.

Popular cathode materials for Li-ion batteries are metal oxides like Lithium Cobalt Oxide ($LiCoO_2$ or LCO), Lithium Manganese Oxide ($LiMn_2O_4$ or LMO), Lithium Nickel Manganese Cobalt Oxide ($Li(Ni,Co,Mn)O_2$ or NMC) and Lithium Nickel Cobalt Aluminum Oxide ($Li(Ni,Co,Al)O_2$ or NCA). Besides these materials, Lithium Iron Phosphate ($LiFePO_4$ or LFP) is also used. However, due to its high price the recovery of cobalt is in general the driver for Li-ion battery recycling activities. Anode materials are usually high-grade graphites, carbon black or Lithium Titanate ($Li_4Ti_5O_{12}$ or LTO). For the current collection, copper and aluminum foils are usually used where the active cathode and anode materials that are mixed with binders (e.g., Polyvinylidene Fluoride PVDF) are coated on. In order to transport the Li-ions between the cathode and anode, an electrolyte, usually a lithium salt such as Lithium Hexafluorophosphate (LiPF$_6$) in an organic solvent, is essential. Besides, a range of additives (e.g., ethylene carbonate or dimethyl carbonate) are applied to optimize the properties of the electrolyte solution. To avoid short-circuiting, polymeric separators are needed.

The choice of cathode material is a trade-off between power, energy, life cycle, cost and safety of the material. Depending on the application certain cathode materials are favored, for example:

- LCO for mobile applications like mobile phones or laptops [Battery University (2017)],
- LMO for power tools, hybrid electric vehicles and e-bikes [Frost & Sullivan (2014); Battery University (2017)],
- NMC for battery and hybrid electric vehicles and e-bikes [Frost & Sullivan (2014); Battery University (2017)],
- LFP for hybrid electric vehicles [Frost & Sullivan (2014)].

With respect to the employed cathode materials, the world market for lithium-ion batteries is segmented, as depicted in Fig. 10.1.

Today, there is no doubt that digitalization electric mobility will become more and more important in the near future. With an increasing global demand for mobile appliances and electric transportation, how will this demand translate to the needed resources, in particular to battery-related resources and their supply?

Lithium: According to the 2017 US Geological Survey [USGS (2017)], battery applications dominate with 39% of the global lithium markets (Fig. 10.2), followed by ceramic and glass applications. Due to large reserves

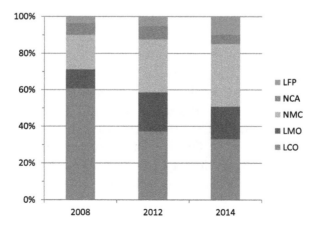

Fig. 10.1 World market size of different cathode materials according to [Heelan et al. (2016)].

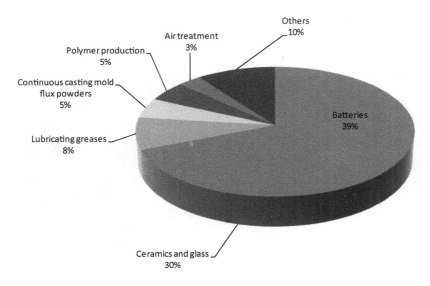

Fig. 10.2 Global applications for lithium in 2016 [USGS (2017)].

the static range of this metal is extraordinary long: Based on the amount of mined lithium in 2015 and its known reserves, this element should last another 444 years. Consequently, lithium is not considered critical in supply. However, due to the expected steep worldwide growth of electric mobility, it is estimated that the demand will considerably exceed the mined production by 2035 [Marscheider-Weidemann et al. (2016)].

Today, about 78% of the global lithium supply comes from only two countries — Australia and Chile. Australia supplies 44.8% of the global mine production, while Chile's share is 33.3% [USGS (2017)]. However, if one were to include reserve supplies as well, Argentina and China can also be added to this list.

Cobalt: The battery industry currently uses 42% of the global cobalt mine production. Other applications include superalloys, hard materials or catalysts (see Fig. 10.3). 98 percent of the world's cobalt supply is mined as a by-product of mainly copper (61%) and nickel (37%) production [Al Barazi et al. (2016)], mostly in the Democratic Republic (DR) of Congo, e.g., in the Tenke Fungurume mine, one of the largest known source for cobalt from deposits. On the other side, China is the unchallenged market leader for the production of refined cobalt [Al Barazi et al. (2016)].

With the forecasted increase of electric mobility, companies have to be aware where they are sourcing lithium and cobalt from. According to Stefan Sabo-Walsh from Verisk Maplecroft, the cobalt sourced in DR of Congo

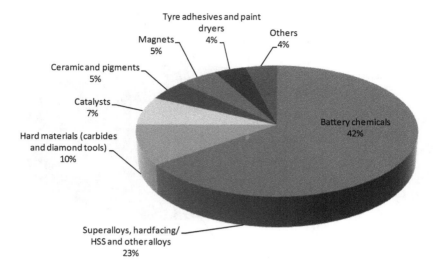

Fig. 10.3 Global applications for cobalt in 2015 [Al Barazi et al. (2016)].

bears the stigma of child labor, human rights abuses, trafficking and health and safety issues, while lithium production issues relate more to weakened land and water rights of the local population [Sabo-Walsh (2017)].

Phosphorous: Phosphorous is an essential element for plants and their growth. With the help of fertilizers, feed and food production could be increased to feed the growing global population. In summary, mineral fertilizers are the major application for phosphorous-based compounds. Besides they are also used as detergents (e.g., for cleaning purposes), additives (e.g., in cement or paint) or chemicals [PFA (2017)]. The latter comprise also the phosphorous-containing compounds LFO or $LiPF_6$, used in Li-ion batteries.

In general, phosphorous is applied in dissipative uses, meaning that it is often lost after being used, usually into waste or ground water. To counteract these losses, phosphorous is increasingly being recovered from sewage water or crop residues.

According to a European study [BIO by Deloitte (2015)], the worldwide extraction of phosphorous in phosphate rock was 37,000 kt in 2012. The largest producers of phosphate rock are China, the United States, Morocco, Peru and Russia. About two-third of the collected end of life products are sent for recycling. However, a significant part of phosphorus, namely the remaining third, is stocked in landfills. Furthermore, phosphorous and phosphate rock were added to the European list of critical raw materials [CRM (2017)].

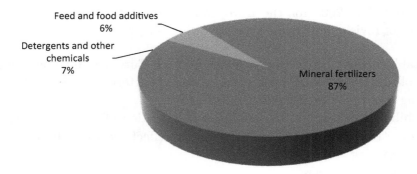

Fig. 10.4 Shares of finished products containing phosphorus used in the EU (taking into account exports and imports of products) in 2012 [BIO by Deloitte (2015)].

10.2 Motivation for Battery Recycling and Legal Framework in Europe

According to the lithium-ion battery cost breakdown performed by Frost & Sullivan (2017) for electric vehicles, it becomes evident that half of the price is that of the battery cell itself. However, more important is the fact that 22.5% of the total costs are material costs. In particular, 13% of the total costs relate to the cathode material being equivalent to almost 58% of the material costs. This results in the high prices for cobalt and its compounds.

In the first commercialized Li-ion batteries, LCO was employed as a cathode material with a high cobalt content of 60 wt%. To decrease the costs, R&D activities concentrate on decreasing the cobalt content in the cathode material, amongst other efforts. The development of alternative cathode materials, such as the ternary metal oxides NCA or NMC, resulted in decreased cobalt content of 10 wt% or 20–30 wt%, respectively.

With respect to the end of the value chain, the collection and recycling of end of life Li-ion batteries is also mainly motivated by the recovery of the valuable element cobalt. The decrease of cobalt content in the cathode materials is also evident in the recycling streams. As reported by Heelan et al. (2016), the amount of cobalt recovered in 2012 in WPI[1] recycling centers amounted to 90% of all recovered elements (Co, Ni, Mn), while it dropped to 60% in 2014. Simultaneously, the amount of nickel and manganese increased from 6 to 25% and 4 to 15%, respectively.

In order to enforce the recovery and recycling of batteries, the Batteries Directive (Directive 2006/66/EC on batteries and accumulators [EU (2006)])

---

[1] Worcester Polytechnic Institute, Worcester (US).

was adopted in 2006 for the European Union and has been subject to a number of revisions. It defines the following targets:

- Establishment of collection schemes for each European member state,
- A collection rate of 25% for waste portable batteries by September 2012, rising to 45% by September 2016,
- Prohibiting of disposal by landfill or incineration of waste industrial and automotive batteries and accumulators, in effect setting a 100% collection and recycling target; and
- Recycling efficiencies:
  - 65% by average weight of lead acid batteries,
  - 75% by average weight of NiCd batteries and
  - 50% by average weight of other waste batteries and accumulators.

In Germany, the Foundation Common Collection System of Batteries (GRS Batterien) is a common collection system founded in 1998 by leading battery producers and the German Electrical and Electronics Industry Association (ZVEI), to ensure nationwide collection, sorting and recycling of used batteries [GRS (2017)]. It is the largest common collection scheme in Europe. Manufacturers and importers use GRS Batterien's services and pay a disposal fee if they put their products on the German market. In 2016, 15.964 tons (t) of batteries have been collected, and it is equivalent to a collection rate of 46.3%. Li-ion batteries amounted to 860 t and primary lithium batteries, 160 t, thus translating to 6.2% of all collected batteries [GRS (2017b)]. With regard to the increasing use of Li-based batteries, the percentage increase in the collection amounted to 31% in 2015.

10.3 Available Recycling Technologies

Due to the wide range of batteries that exist and the varying metals that they are made of, specific recycling processes have been developed for each battery type. Prior to recycling, the first step is to sort the batteries into groups by type. Where batteries are not collected separately, they enter the municipal waste stream and are either landfilled or incinerated. Since Li-ion batteries are the battery type with the strongest growing market, this chapter focuses on the recycling of these products.

Various approaches have been developed to recover the most valuable materials from the complex material mix that the modern Li-ion battery modules consist of. These approaches are usually a combination of mechanical treatments and pyrometallurgical and/or hydrometallurgical steps, which are primarily aimed at regaining metals or oxides.

10.3.1 Pre-processing treatments

The recycling of Li-ion batteries is more complicated compared to lead–acid or NiMH recycling. Li-ion battery cells contain a bigger variety of materials. The powder-shaped active materials are coated onto metal foils, while lead–acid batteries contain a small number of large lead plates and consist of one cell. A Li-ion pack consists of several individual cells, e.g., the battery pack in an electric vehicle from Tesla includes about 5000 cells [Gaines (2014)].

Since Li-ion batteries can explode due to their compact design, the reactive lithium, the short distance of the anode and cathode and the associated risk of short circuits by means of impurities or increased temperature, they have to be discharged prior to further treatment and handled with care. Some crushing processes are carried out under cryogenic conditions to reduce the reactivity of lithium [Laucournet et al. (2014)]. Figure 10.5 depicts a scheme from Laucournet et al. (2014) for the pre-processing of Li-ion battery packs and possible recycling strategies.

i. Sorting methods
As each battery's chemistry tends to require a different recycling process, all waste batteries must be accurately sorted to avoid cross-contamination down-stream at the recycling plant. The German Foundation Common Collection System of Batteries (GRS Batterien) uses two sorting methods [GRS (2017c)]:

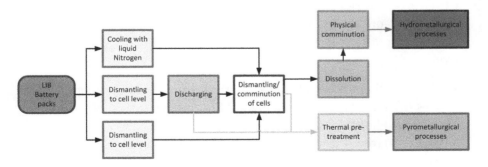

Fig. 10.5 Pre-processing of Li-ion battery packs and basically possible recycling strategies [Laucournet et al. (2014)].

Electromagnetic process:
An electromagnetic sensor identifies magnetic round cell systems through a magnetic field, which is disturbed depending on the electrochemical system of the cell. The cell is detected by an induced change in voltage. With this method a sorting purity of a minimum 98% is reached.

X-ray process:
After sorting the round cells by size, the grey gradation of an image obtained after x-irradiation is measured to distinguish between the different applied materials. Here, the purity is also more than 98%.

ii. Disassembling

In some cases a disassembly of battery packs to a cell level is mandatory (or at least favorable). For example, in hydrometallurgical processes impurities like plastics, binder, etc., should be avoided. In the later described electrohydraulic fragmentation process, the plastic packaging parts have a disruptive effect and can extend or prevent the process. Wegener et al. (2015) made investigations on robotic assisted dismantling, with a focus on camera-assisted detection of bolts and their removal. The method is automatic but time consuming. It needs to be improved in terms of accuracy and speed.

iii. Shredding

For pyro- and hydrometallurgical processes the batteries or battery packages have to be roughly shredded. Basically, the method should chop, crush, and/or break the batteries into small pieces for further treatment. The individual machines vary in detail. However, in general, the separation of the different material fractions is required to facilitate optimum separation and sorting.

iv. Physical separation and sorting

Depending on the recycling process, sorting of the different fraction is necessary after mechanical, chemical and thermal treatment, respectively.

- **Sieving**: Separating of mixed materials by utilizing different particle sizes to divide them into rough and fine fractions; this method is simple and cost-effective.
- **Windsifting**: Comprises various procedures using gravity and centrifugal forces under air-flow, e.g., in a zigzag-sifter, pieces of plastic foil can be separated from heavier materials.

- **Magnetic separation**: Magnetic materials, e.g., steel housings, are separated by means of electromagnets or permanent magnets.
- **Electrostatic separation**: Electrostatic materials, such as plastics like PVC and PET, can be separated using the friction in a rotating drum; negative and positive charges get separated in the plastic parts and the materials can be separated according to their static charge.
- **Eddy current process**: After all ferrous metals have been removed, the non-ferrous metals can be separated from other materials by eddy currents generated in the metal parts through rotating magnets. According to Lenz's rule, the eddy currents generate magnetic fields in the metal parts, leading to magnetic repulsion. As a result the metal parts drop out.
- **Flotation process**: Fine-grained particles in a fluid can be separated from each other by means of their wettability; gas bubbles easily accumulate at hydrophobic particles, thus they can be floated to the surface of the fluid to be skimmed off.

v. Thermal pre-treatment

For several reasons a thermal pre-treatment is reasonable: Optimum safety for the battery recycling process, avoiding the emission of electrolyte material, and optimized recovery conditions for electrode materials by removing the binder. At temperatures below 380°C, the binder PVDF decomposes and oxidation of carbon takes place. Below 350°C, the decomposition of some hazardous solvents also starts. At higher temperatures and controlled atmosphere, ashing or combustion of plastics takes place.

10.3.2 *Pyro- and hydrometallurgy for extraction*

When batteries within a particular group are collected separately for recycling to avoid complex material mixtures, there is a variety of methods to adequately recover the valuable materials. There are two basic types of processes — pyrometallurgical (where batteries are placed in a furnace and treated thermally) and hydrometallurgical (where batteries are treated chemically to separate elements and compounds). Most processes designed for specific battery chemistry are intolerant towards contamination by other chemistries. Thus, it is vital that batteries are sorted and supplied to the recycling plants within the specified tolerances of that plant.

Pyrometallurgical processes

In conventional recycling processes, batteries are either melted directly or — for the purpose of homogenization — mechanically comminuted in an upstream process step and then thermally processed. With conventional, mechanical fragmentation, selective removal of individual components is only possible to a certain extent, thus preventing the removal of a clean plastic or metal fraction. Low-concentration recyclables are not recovered in the heat treatment but lost in the slag. The plastic fraction is recycled to energy in the melting process.

In pyrometallurgical processes, heating and thermal energy is utilized to start reactions that cause the transformation of materials. At low temperatures only phase transitions or structural changes take place, whereas at higher temperatures chemical reactions occur. Such a process is often referred to as calcination or pyrolysis (depending on the temperature). The materials start to melt, if the temperature is sufficient (as an indication: $T > 700°C$). In the case of melting batteries, a metallic fraction, a slag and gases are generated. The gas phase consists of volatile reaction and degradation products of organic components, and volatile metals such as lithium, zinc and mercury (provided they are contained in the batteries), which is not the case for the mentioned elements for Li-ion batteries except for lithium).

Pyrometallurgy is used in industrial processes to extract metals and alloys. These processes are safe, profitable, fast and simple to implement, but not selective, i.e., the recovery of a specific metal is almost impossible. On the other hand, organic components and the electrolyte contribute to the combustion process, thus making it more energy efficient. Lithium and aluminum end up in the slag, where they have to be extracted by hydrometallurgical processes.

Additional drawbacks are the high-energy demand and the need for cost-expensive gas purification systems [Laucournet et al. (2014)].

Hydrometallurgical processes

The aim of hydrometallurgical processes is to dissolve elements or materials in acids, alkalis, saline solution or other solvents, in order to extract them in the following process steps. The key benefits are the chemical selectivity and high extraction, and recycling efficiency, respectively. Furthermore, these processes do not need high temperatures and the emission of gases is low.

As a drawback, the need for a high volume of chemicals has to be mentioned. Since some of the reactions take time, the recycling rate per time unit is less efficient as compared to pyrometallurgical processes [Laucournet et al. (2014)].

The term hydrometallurgy covers the following chemical processes:

- Dissolution,
- Solvent extraction,
- Leaching and bioleaching,
- Chemical precipitation,
- Ion exchange, and
- Electro-chemical processes.

For recycling purpose, **leaching** is a common technology to selectively extract soluble elements from solids by means of a liquid media. For a Li-ion battery, recycling acid leaching is commonly used, in particular, strong acids such as sulfuric acid, hydrochloric acid and nitric acid.

Dissolved ions can be extracted by **precipitation** reactions by adding a chemical reagent that selectively reacts with this ion. A solid compound is formed that has to be separated from the solution by filtering or centrifugation. Usually, the substance has to be washed several times afterwards.

Dissolved ions can also be extracted by putting them together with the dispersed metal into a column filled with an **ion exchange** medium, where the ions get absorbed. In a subsequent washing procedure the ions/elements are recovered.

The basis of the **solvent extraction** is an immiscible two-phase system of an organic and an aqueous solution. Common reagents are Di-(2-ethylhexyl)-phosphoric acid (D2EHPA), Bis-(2,4,4-tri-methyl-pentyl)-phosphoric acid (Cyanex 272), Trioctylamin (TOA), Diethylhexyl phosphoric acid (DEHPA) or 2-Ethylhexyl-phosphoric acid-mono-2-ethylhexyl ester (PC-88A) [Laucournet et al. (2014)]. The main benefit of this method is its high selectivity, which allows for the extraction of individual ions (such as Li-ions). The extracted compounds are of higher quality as compared to the compounds extracted from precipitation reactions. However, its disadvantages are its high costs and the toxicity of the applied chemicals (such as benzene and toluene).

Electrolysis is an **electrochemical process** where an ion-permeable membrane is located between an anode and a cathode inside a cell. At the cathode side, metal ions accumulate until the limit of solubility is reached and metal hydroxide precipitates.

10.4 Electrohydraulic Fragmentation, an Innovative Recycling Process for Battery Recycling

In cooperation with the fragmentation equipment manufacturer ImpulsTec, the Fraunhofer Project Group for Materials Recycling and Resource Strategies IWKS has developed an innovative recycling process to improve the recycling of lithium-ion batteries. This method is based on electrohydraulic fragmentation (EHF). The first pilot system for the industrial application of this technology was installed at the Project Group in Alzenau, Germany.

The operational principle is comparatively simple. It basically consists of a reactor filled with water and electrodes that can be inserted (see Fig. 10.6). The simple design of the plant enables easy upscaling and adaption to specific customer needs.

Crushed batteries are brought into the reactor, which is filled with water. Electrodes are immersed into the water and the reactor is closed. Then an electric arc is ignited between the electrodes by electric discharge at voltages of up to 50 kV. During the process shockwaves are created which propagate through the water and hit the batteries uniformly. In contrast to the purely mechanical crushing, the batteries do not get simply crushed, but are separated at their phase boundaries by the shockwaves. The degree of separation can be selectively adjusted by the process parameters and the number of discharges to be generated by the electrodes. The high selectivity of the fragmentation is caused by different properties of the batteries or other end of life products like interfaced materials. In particular, reflections and interferences of the shockwaves at interfaces of materials with different

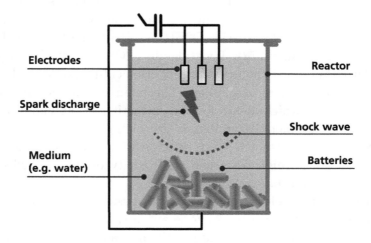

Fig. 10.6 Schematic functional principle of electrohydraulic fragmentation.

Fig. 10.7 Battery materials after EHF and separation into fractions; (*top*) battery cells, (*left*) metal parts, (*bottom*) black mass (mainly anode and cathode materials) and (*right*) plastics.

acoustic characteristics cause specific damage to the materials at their mechanical weak points. This allows them to be separated into different material classes like metals, ceramics, glass or polymers.

In addition, the comminution in water is beneficious for the safety of the process. Hazardous dusts or dangerous reaction products can be captured and passivated in the water (optional with additives). Furthermore, the "contact free fragmentation" (i.e., no strong contact with hard and abrasive grinding tools, only loose contact to the reactor) reduces possible contaminations, for instance, with iron. Thus, the process is well suited for the creation of materials fractions with a high cleanliness. After the shockwave treatment the materials can be sorted by simple physical separation processes like sieving, filtering, magnetic separation or sensor based separation. Figure 10.7 presents the fractions obtained after sorting. The processes are completely free from critical chemicals, which result in economic and ecological benefits. The facilitated sorting of materials fractions will lead to significantly improved efficiencies of subsequent metallurgical treatment processes. The reason is that the reduced volumes of highly concentrated materials only require a reduced amount of chemicals (or of process energy). These materials can be efficiently treated and recycled into new products. In addition, the recovery of materials in high purity enables an advanced functional recycling: Functional compounds (such as electrode materials) have the potential to be reused for new products without the need of complex synthesis routes starting from the metals. This is particularly useful for the treatment of production residues.

10.5 Outlook

It is common knowledge today that global megatrends like digitalization, connectivity and e-mobility will influence our lives in the near future. These developments require specific building block elements such as lithium and cobalt that are relevant for state of the art batteries. The question of securing supplies for these resources is especially important for nations that do not have direct access to them in their own country. Thus, recycling of high-tech applications, and also batteries, is an important means to not only recover such valuable resources, but to also prevent harmful substances from entering the natural environment.

Bibliography

Al Barazi, S., Elsner, H., Kärner, K., Liedtke, M., Schmidt, M., Schmitz, M., and Szurlies, M. (2016). *Mineralische Rohstoffe in Australien — Investitions- und Lieferpotenziale*, Deutsche Rohstoffagentur (DERA). (Ed.) Bundesanstalt für Geowissenschaften und Rohstoffe (BGR). DERA, Berlin.

Battery University (2017). http://batteryuniversity.com/learn/article/types_of_lithium_ion. Date accessed: 25 October 2017.

BIO by Deloitte (2015). Study on Data for a raw material system analysis: Roadmap and test of the fully operational MSA for raw materials. Prepared for the European Commission, DG GROW.

CRM (2017). List of critical raw materials, https://ec.europa.eu/growth/sectors/raw-materials/specific-interest/critical_en.

EU (2006). Directive 2006/66/EC of the European Parliament and of the Council on batteries and accumulators and waste batteries and accumulators and repealing, Directive 91/157/EEC, 6 September 2006.

Frost & Sullivan (2010). Vishal Sapru, Analyzing the global battery market, *Battery Power* **14**, 4, pp. 4–8.

Frost & Sullivan (2014). Global analysis of the electric vehicles lithium-ion batteries chemicals and materials market, Market Engineering M98E-98.

Frost & Sullivan (2017). Global electric vehicle market outlook, 2017, Strategic Insight MCC9-18.

Gaines (2014). Sustainable materials and technologies, **1–2**, pp. 2–7.

GRS (2017). Stiftung GRS batterien (in German), http://www.grs-batterien.com. Date accessed: 25 October 2017.

GRS (2017b). Stiftung GRS batterien (in German); Erfolgskontrolle 2016: GRS batterien steigert erneut sammelmenge, Press release, http://www.grs-batterien.de/grs-batterien/aktuelles/singleansicht/article/erfolgskontrolle-2016.html, 30 April 2017. Date accessed: 25 October 2017.

Heelan, J., Gratz, E., Zhang, Z., Wang, Q., Chen, M., Apelian, D., and Wang, Y. (2016). Current and prospective Li-ion battery recycling and recovery processes, *Journal of The Minerals, Metals & Materials Society* **68**, 10, pp. 2632–2638.

Laucournet, R., Garin, G., Senechal, E., and Yazicioglu, B. (2014). Report on the ELIBAMA project, Li-ion batteries recycling — The batteries end of life, https://elibama.files.wordpress.com/v-d-batteries-recycling1.pdf. Date accessed: 25 October 2017.

Marscheider-Weidemann, F., Langkau, S., Hummen, T., Erdmann, L., Tercero Espinoza, L., Angerer, G., Marwede, M., and Benecke, S. (2016). Rohstoffe für Zukunftstechnologien 2016 — DERA Rohstoffinformationen **28**, 353 S., Berlin.

PFA (2017). Phosphate forum of the Americas, http://phosphatesfacts.org/uses-applications/. Date accessed: 29 October 2017.

Sabo-Walsh, S. (2017). The hidden risks of batteries: Child labor, modern slavery, and weakened land and water rights, Greentech Media, 20 March 2017, https://www.greentechmedia.com/articles/read/green-battery-revolution-powering-social-and-environmental-risks#gs.sRBCoVU. Date accessed: 27 October 2017.

USGS (2017). Mineral commodity summaries: Lithium, https://minerals.usgs.gov/minerals/pubs/commodity/lithium/. Date accessed: 25 October 2017.

Wegener, K., Chen, W. H., Dietrich, F., Dröder, K., and Kara, S. (2015). Robot assisted disassembly for the recycling of electric vehicle batteries, *Procedia CIRP* **29**, pp. 716–721.

Chapter 11

Power-to-X Conversion Technologies

Friedrich-Wilhelm Speckmann and Kai Peter Birke

11.1 Definition of Power-to-X

Power-to-X (PtX) describes the conversion of electricity as a primary energy carrier to a secondary energy carrier like heat, cold, fuel or basic materials for the chemial industry. It is a collective term for Power-to-Gas (PtG), Power-to-Liquid (PtL), Power-to-Solid (PtS), Power-to-Fuel, Power-to-Chemicals and Power-to-Heat. One of the key components is hydrogen, which can be generated from renewable energy via electrolysis or from fossil fuels, mainly natural gas. Hydrogen in conjunction with a carbon source, mostly carbon monoxide or carbon dioxide, can be converted into various products, as shown in Fig. 11.1. These synthesized materials are, for example, fuels, fertilizers or chemical products.

The multitude of possible products makes PtX a versatile and promising technology for the future. Nowadays, different technologies have varying readiness levels and therefore, face diverse challenges in a wide range of potentials that are highlighted in this chapter.

11.2 Potential of Cross-Sectoral Applications

Power-to-X has the potential to be a versatile, cross-sectoral solution for energy transition from renewable generated electricity, into various energy consumption areas. The technology for some of the conversion possibilities is already mature and ready for use in the specific sectors. However, it is not often commercially viable at this point of time. On one hand, an increase in the interconnection of the different sectors and a bigger share of renewable energy can improve this situation. On the other hand, energy storage is mandatory for a cross-sectoral solution, as displayed in Fig. 11.2.

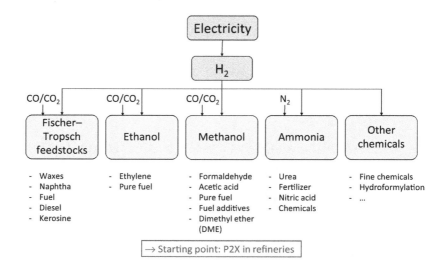

Fig. 11.1 Power-to-X conversion chains from electricity into value-added materials.

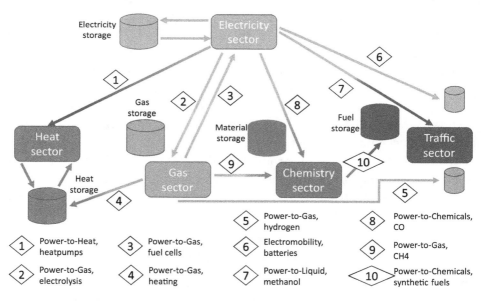

Fig. 11.2 Overview of cross-sectoral interaction of various Power-to-X processes and required technologies.

The flow chart shows that electricity, which has been generated from fossil fuels or renewable energy sources like wind- or solar-power, serves as the starting energy form for the PtX system. The electricity can be stored in batteries or be converted into energy carriers for the heat-, gas-, chemistry- and traffic-sector. Especially, the gas sector has versatile conversion options into further areas of application. The gas, e.g., hydrogen or methane from electrolysis or methanation processes, can be processed in a block-heat and power plant or can be used for mobile applications via fuel cells. Furthermore, it is possible to subsequently synthesize the gas and transform it into liquid synthetic fuels.

Another area of application for storage systems is grid-based. The use of energy storage systems (EES) in the electrical grid is often argued to be able, to reduce the necessary electrical grid expansion. This might be the case in some instances, where the systems are used for grid stabilization and not for economic purposes. But in general, this is not a valid argument, because most EESs are operated as local intermediate storage systems and cannot solve the regional imbalance in electrical energy generation. Only PtX technologies in particular have a special status and are not limited to the area of generation. As an example, PtG systems can be connected to the existing gas infrastructure and feed the synthesized methane, or even a limited amount of hydrogen [Specht *et al.* (2009)] directly into the gas grid. Therefore, it is possible to conduct a simultaneous shifting of electrical energy in a temporal and regional fashion. Figure 11.3 shows different energy storage technologies grouped together by their capability of discharge duration as well as storage capacity.

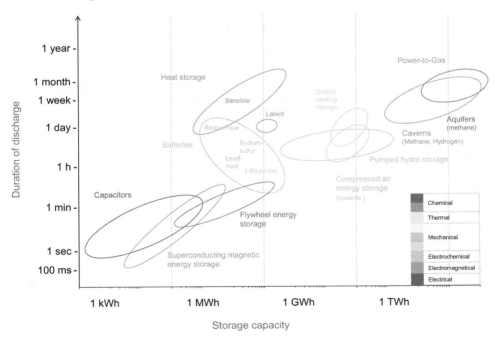

Fig. 11.3 Storage technologies grouped by their duration of discharge and capacity.

Besides the technical characterization of EESs, the technologies can be grouped by further aspects. Firstly, the field of application. Is the storage integrated into interconnections or an island network? Secondly, on which voltage level does it operate? Thirdly, to which electricity market does it belong? Is the storage used for power trading, standby energy or optimizing installation consumption in households and industry?

A storage technology can have various of these characteristics. Increased flexibility of storage systems strengthens their economic viability. The primary goal of every storage is the balance of energy surpluses and shortages. For the electric grid, this equals times of high energy generation and times of increased load that result in different prices in the electricity market. Trading based on the differences in price is called arbitrage business and it is, together with the trade of standby energy, a financially promising area for EES technologies. Therefore, storages generate the highest income at times of peak demand and at very low levels of energy generation. The economical viability depends on the frequency of this spread between demand and generation. Since the Renewable Energy Act of 2007, electric energy from renewable sources has to be prioritized over electricity from fossil fuels [Pachauri and Reisinger (2007)]. Conventional power plants, for example, coal-based facilities, are slow in regulating their power output. Thus, they are mainly used for base load operation. The decrease of the power output is therefore, often not economically viable, even in times of overall high generation and low electricity prices. In extreme cases this leads towards negative electricity prices. Storages with high durations of discharge (cf. Fig. 11.3) could be charged during this periods in order to increase their profitability.

Besides the arbitrage business, energy storages are key technologies for providing ancillary grid services and are necessary for a reliable, safe grid operation. It is essential that the voltage level and its frequency in the respective grids are regulated. The national Grid Code restricts the allowed conditions in the electricity grids, e.g., the Union for the Coordination of the Transmission of Energy (UCTE) limits the voltage to be within the range of 0.95 p.u. and 1.05 p.u. of the nominal value. This frequency is only allowed to diverge 0.05 Hz from the European norm of 50 Hz [Entsoe (2017)].

Grid stabilization cannot only be achieved by the integration of PtG systems, but also by the use of stationary battery storage. Large-scale stationary battery storage has been under development for more than two decades and many large battery demonstration projects have been realized. A variety of electric utility grid applications, in conjunction with renewable energy sources, have already been tested. There are two main fields of applications: energy applications and power applications. Energy applications involve storage systems that discharge over periods of hours, with correspondingly long charging periods. Power applications involve comparatively short periods of discharge in the range of seconds to minutes, short recharging periods and often run many cycles per day. Though battery categories have been developed and tested for these purposes, only lead-acid, sodium/sulfur, nickel/cadmium, vanadium-redox flow and lithium-ion batteries have been commercially employed.

Lithium-ion batteries that do not fulfill the demands of mobile applications (50–80% of initial capacity, 200–400% of initial internal resistance) anymore, can also be used for stationary storage systems and have lower investment costs. Simultaneously, their lifetime is lower and a modified battery management system (BMS) has to be installed for their new purpose of grid services. The use of recycled batteries from the electro-mobility sector is called second-life application. What the above-mentioned battery types have in common is that their energy density, capacity as well as power output, cannot be independently scaled. Furthermore, does every stationary battery storage system require a BMS and shorter maintenance intervals than PtG systems, since every battery cell can behave differently and has to be balanced. The lifetime also depends on the amount of charge and discharge cycles, as well as its calendaric degradation.

Problematic areas for battery technologies are in space and aircraft applications where the gravimetric energy density matters the most. Conventional lithium-ion battery modules have a practical energy density of 100–150 Wh/kg, because the required module housing increases the absolute weight and thus, decreases the specific capacity of the entire system (cf. Chap. 3). On the contrary, PtG storage systems are flexible in size and use lightweight gases like H_2 with a remarkable high gravimetric energy density of 33.3 kWh/kg [Bossel (2003)]. These gases are used in conjunction with fuel cells and require a pressure tank which increases the total weight of the whole system, but that still results in a higher gravimetric energy density than batteries.

The legal demands for PtG systems are very specific. Not only are the renewable sources of electricity relevant, but so too are the final gas quality. As an example, the German "Energiewirtschaftsgesetz" (§3, 2011) states that synthetic natural gas (SNG) has to be obtained from electrolyzers fed by mainly renewable energy, in order to be legally fed to the national gas grid. The system regulation from the "Deutscher Verein des Gas- und Wasserfaches e.V." states that the gas has to be more than 95% methane

(CH_4) with a minimum heating value of $8.4\,\text{kWh/m}^3$. The remaining 5% are mainly hydrogen (H_2), carbon monoxide (CO) and carbon dioxide (CO_2). The amount of hydrogen is especially relevant, because of the different application types connected to the gas grid. For example, gas power plants are equipped with gas-turbines that cannot handle a H_2 amount of over 2% and therefore, the total percentage of methane, in the German national grid, must not be reduced to under 98% [Zapf (2017)].

11.3 Power-to-X as a Primary Battery

There are two major types of battery cell technologies. On one hand, a primary cell that is regarded as dead after the reactants are fully used up by the discharge process. On the other hand, a secondary cell, capable of being recharged after the reactants have been used up. The spontaneous chemical reaction can be inverted by passing an electrical current through the cell in the inversed direction of the discharge process. Therefore, a secondary battery can be considered as an electrical energy storage [Vincent and Scrosati (2014)].

The related subject of fuel cells can also be regarded as primary batteries, where the anodic reactant, in most cases a gas, is stored externally. It is further possible to supply the fuel cell with a constant gas feed and thus, create a kind of hybrid cell that is considered an energy storage. There are different relevant fuel concepts used for PtX applications. In general, most of these systems are relatively simple compared to large-scale battery systems. This can lead to advantages of fuel cells in special areas of use, and therefore, it is necessary to highlight this technology as a hybrid cell. Especially, the possibility for long term storage and almost no limitations for the scalability of such systems result in a promising technology for the integration of renewable energy sources into the electric grid. Hydrogen from electrolysis systems as well as natural gas from methanation can be stored for months. This enables a seasonal balancing of energy generation.

11.4 Power-to-Gas

The aim of the Power-to-Gas (PtG) is to convert electricity from renewable energy sources to gaseous energy carriers, mainly hydrogen or methane. These renewable gases can be transported in the existing gas infrastructure and be used in a wide range of applications. They can be generated in times of high energy generation and reconverted when in demand. Likewise, the direct use of hydrogen, for instance in the chemical or mobility sector, is

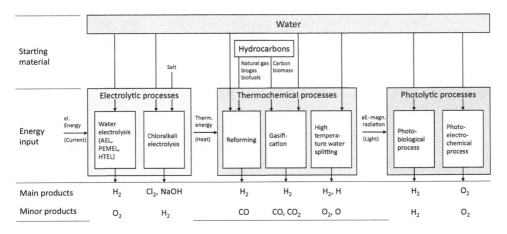

Fig. 11.4 Hydrogen production processes and technologies.

possible. The PtG technology can reduce CO_2 emissions in different sectors of consumption by replacing fossil fuels with synthesized renewable gas. Furthermore, it is possible to filter the required CO_2 for the methanation process directly from the atmosphere. Thus, leading to a CO_2 neutral balance for the entire process, from the generation of electricity to the combustion of the renewable methane gas.

The technology is mature and ready for use. However, until it is completely ready for the market, in particular with regard to economic use, a few central regulatory parameters need adjusting. The focus lies on the improvement of electrolysis and methanation processes, including the required power electronics.

11.4.1 Hydrogen generation

Hydrogen is the key element for the PtG technology and can be generated via various processes. The main generation processes are electrolytic, thermochemical as well as photolytic. Figure 11.4 displays the requirements, outputs and connections of these systems for the generation of hydrogen.

The shown processes all demand water and some kind of energy as their primary feedstock. Depending on the kind of process, additional input is required, for example, the addition of biomass for a gasification system.

11.4.2 Electrolytic hydrogen generation

The electrolysis of water, in order to generate hydrogen, is one of the core processes of the PtG concept. An electrolyzer splits water into hydrogen and oxygen using electricity. Four technologies are the most relevant in this context — the alkaline water electrolysis (AEL), polymer electrolyte membrane electrolysis (PEMEL), high-temperature electrolysis (HTEL) and the chloralkali electrolysis (CAEL).

The AEL has a long history in the chemical industry and is a proven process, even in large scales. This electrolyzer technology is based on an aqueous alkaline electrolyte, most commonly potassium hydroxide (KOH) with concentrations between 25–35 wt% [Ulleberg (2003)], in order to increase the ionic conductivity. Hydrogen gas evolves at the cathode and the resulting hydroxide anions (OH^-) travel through the gas-tight membrane to the anode within the electric field. The reactions at the electrodes of an alkaline electrolyzer can be separated in the anode- (see Eq. (11.1)) and the cathode-reaction (see Eq. (11.2)).

$$4OH^-(aq) \longrightarrow O_2(g) + 2H_2O(l) + 4e^-, \tag{11.1}$$

$$4H_2O(l) + 4e^- \longrightarrow 2H_2(g) + 4OH^-(aq). \tag{11.2}$$

It is a mature technology that has been available on the market for the last 50 years [Kreuter and Hofmann (1998)]. The electrolyzer itself is relatively simple and can be operated at atmospheric- or at higher pressure levels for a lifetime of up to 30 years. Furthermore, they use no platinum group metals (PGM) catalysts, commonly in the form of Ni-based electrodes, which result in low initial investment costs. Additionally, AEL is scalable over a wide range of installed power. The major shortcoming of conventional alkaline systems is their low operating current density [Marini et al. (2012)].

The PEMEL technology makes use of an ionic conducting membrane that allows the transport of H^+ ions from the anode to the cathode, where they recombine with the electrons to form H_2. Due to this process, the technology is known as the proton exchange membrane electrolyzer [Barbir (2005)], with the following electrochemical reactions at the anode (see Eq. (11.3)) and the cathode (see Eq. (11.4)):

$$2H_2O(l) \longrightarrow O_2(g) + 4H^+(aq) + 4e^-, \tag{11.3}$$

$$4H^+(aq) + 4e^- \longrightarrow 2H_2(g). \tag{11.4}$$

Most commonly used types of membranes consist of gas-tight Nafion with a cross-linked structure and have a strongly acid characteristic because of the functional sulfonic acid groups ($-SO_3H$), which enable their proton conducting ability [Larminie and Dicks (2003)]. Despite their efficiency and high current density [Ulleberg (2003)], there are drawbacks. The lifetime strongly depends on the operating temperature — a high temperature results in increased efficiency but decreases the durability. Therefore, lifetimes are in the range of five to ten years [Kreuter and Hofmann (1998)]. More importantly, the membranes degrade over time and have to be a focus of maintenance. Another problem is the high investment costs due to the expensive membrane and the use of noble metals as catalysts, such as Pt, Ru and Ir [Marini et al. (2012)].

In HTEL systems a part of the energy required for splitting water is provided by the temperature, usually in a range between 700–1000°C. Thus, lowering the minimum cell voltage down to 1 V (which is possible due to thermodynamic reasons) will enable increased current efficiency rates. The HTEL is based on the reverse reaction of the solid oxide fuel cell. Both half-cells are separated by a ceramic solid oxygen ion (O^{2-}) conductor with the electrodes on each side of the surface. The cathode side is supplied with super-heated water vapor, which reacts with two electrons to hydrogen and O^{2-}-ions, as shown in Eq. (11.5). Hydrogen can be extracted and the ions diffuse to the anode side, where they react with oxygen (cf. Eq. (11.6)).

$$2H_2O(g) + 2e^- \longrightarrow H_2(g) + O^{2-}, \tag{11.5}$$

$$O^{2-} \longrightarrow 1/2O_2(g) + 2e^-. \tag{11.6}$$

The advantages of HTLEs are based on their high operational temperature which favors the reaction kinetics. At 2500°C no electrical input to split water would be required because water breaks down due to thermolysis. Furthermore, energy in the form of heat will cost less than electricity. Therefore, an operation with waste heat seems economical reasonable. However, the greatest disadvantage is that the high operating temperature will result in long start-up times and break-in times. The high operating temperature leads to mechanical compatibility issues, such as thermal expansion mismatch. Chemical stability problems, such as the diffusion between layers of material in the cell are common. Thus, making HTEL systems a niche application at the research phase [Goetz et al. (2016)].

In contrast to the above-mentioned electrolyzers, the CAEL does not focus on the generation of hydrogen in general. It is a process that decomposes a salt solution in water, simultaneously producing chlorine (Cl_2), caustic soda ($NaOH$) and H_2. Therefore, this technology has special potential in a future PtX environment with cross-sectoral energy conversion. Several of these processes are available globally; the most important routes are the following: ion exchange membrane process, mercury cell process and diaphragm process. The geographical distribution of these three technologies varies significantly and depends on the demand in that region.

Table 11.1 Characteristics of commonly used electrolysis systems [(l) = liquid, (s) = solid].

	AEL	PEMEL	HTEL	CAEL
Electrolyte	KOH (l)	Membrane (s)	Ceramic (s)	Membrane (s)
Temperature [°C]	40–90	20–100	700–1000	85–90
Pressure [bar]	1–30	30–50	~30	0–0.25
Charge carrier	OH^-	H^+	O^{2-}	Na^+, Cl^-
Efficiency [%]	62–82	67–82	65–82	94–97
Active cell area in $[m^2]$	0.1–4	0.1–0.75	0.01–0.1	1.5–3.3
Current density $[A/cm^2]$	0.2–0.45	up to 2.5	0.3–3.0	0.15–0.8
Power $[kW]$	5–3400	0.5–160	1–18	1–800
Energy $[kWh/m^3 H_2]$	4.5–7.0	4.5–7.5	$3.2 + 0.6_{therm}$	7.1–7.5

Table 11.1 summarizes the described electrolysis processes and compares their characteristics.

All presented processes have specific advantages for certain applications and can be potentially used in a PtG environment.

11.4.2.1   Thermochemical hydrogen generation

The most common hydrogen generation method is steam reforming with an annual global H_2 generation of 190 Mrd. Nm^3, out of the total amount of 500 Mrd. Nm^3 H_2 [Gardiner (2012)]. Fossil energy carriers with a high amount of hydrogen molecules (e.g., natural gas, methanol) are fed to a reformer. Upon the addition of water vapor, nickel catalysts and heat energy, a hydrogenous gas is formed. Afterwards, a shift-reaction takes place which reduces the amount of CO and water vapor and converts it into CO_2, as shown in Eqs. (11.7) and (11.8):

$$CH_4(g) + H_2O(g) \longrightarrow CO(g) + 3H_2(g), \qquad (11.7)$$

$$CO(g) + H_2O(g) \longrightarrow CO_2(g) + H_2(g). \qquad (11.8)$$

Subsequently, a gas purification reduces the remaining unconverted CO to a minimum. The generation of 1 Nm^3 H_2 requires, based on the size of the reactor, an amount of roughly 0.45 Nm^3 natural gas. Therefore, the efficiency of the whole process is about 80%. The amount of required heat energy strongly depends on the used substrate. Methanol for example has a single bond between the carbon atom and the hydroxide group. This structure is slightly polarized and therefore, only requires around 300°C to be split up. In contrast, natural gas has the same single bond between the carbon atom and the hydrogen, but without the oxygen atom. Thus, the structure is not polarized and requires a temperatures of up to 800°C in order to separate this bond. Benzine consists of several carbon-ion bonds that can only be broken down with a temperature of over 900°C.

11.4.2.2 *Photochemical hydrogen generation*

The possibility of generating H_2 from water splitting, by directly using a source of renewable energy such as solar light can be of particular interest. An essential component for the process is a light-absorbing chromophore, often called photosensizer. The excited state, with its low-energy hole and high-energy electron, can achieve water splitting in two possible ways — water reduction followed by water oxidation or vice versa. In theory, many dyes absorbing visible light (1.5–3.1 eV) have the potential thermodynamic power to perform this photochemical reaction, but there are challenges of a kinetic nature. While the charge separation and recombination steps initiated by light absorption are one-electron processes, the oxidation and reduction of water are slow, multi-electron processes. Therefore, every water splitting photochemical cycle must have multi-electron catalysts, be able to store electrons or positive charges and to deliver them to the substrate via low activation energy processes. Figure 11.5 shows a simplified scheme for a photochemical water splitting system [Ladomenou *et al.* (2015)].

All photochemical systems, however, are different in complexity, and are still far away from economical feasibility in large-scale hydrogen production. This could change if the turnover of the components and stability are improved, thus making such systems more viable in the future.

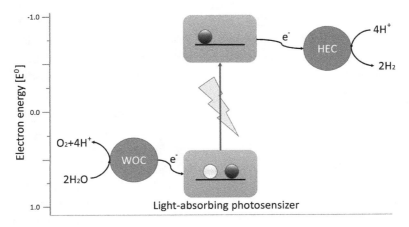

Fig. 11.5 Photochemical hydrogen generation process. WOC = water oxidation catalyst, HEC = hydrogen evolving catalyst.

Furthermore, there are photo-biological hydrogen generation processes based on microorganisms, like green algae or cyanobacteria. They are able to use light in order to split water into H_2O and O_2. The enzyme hydrogenase is able to use the developed H^+-ions, together with the gathered electrons, to form pure hydrogen. The hydrogenase functions as a catalyst in this reaction. Microorganisms can be viewed analogically to plants which use photosynthesis, but with carbon-hydroxides instead of H_2O, as a final product. In comparison to other hydrogen generating methods, this technique is still very slow and inefficient [Wietschel et al. (2015)].

11.4.3 Methanation

One of the great challenges of the 21st century is the conversion of greenhouse gases, mainly CO_2, into value-added chemicals, as well as fuels [Sterner and Stadler (2014)]. Much research has been carried out to develop energy- and cost-efficient technologies. The most common and promising technologies are highlighted in this section.

11.4.3.1 Catalytic/chemical methanation

The Sabatier process is the conversion from hydrogen into SNG (synthetic natural gas), as shown in Eq. (11.9). All reaction participants are in their gaseous state.

$$CO_2 + 4H_2 \longrightarrow CH_4 + 2H_2O \longrightarrow CH_4 + O_2 + 2H_2. \tag{11.9}$$

The entire process consists of the water-gas-shift reaction in combination with a CO methanation. The right hand side of Eq. (11.9) involves a water electrolysis process. The reaction is strongly exothermic. At higher temperatures, the reaction equilibrium is shifted to the educt side and less SNG is formed. Therefore, the heat dissipation is of the highest importance. Reactor technologies are categorized into two- or three-phase systems.

The first group consists of fluidized-bed and fix-bed reactors with a solid catalyst. Three-phase systems, on the other hand, are bubble column reactors. Their catalyst is also solid, but is dissipated in a mineral oil and has the characteristics of a fluid. The fixed-bed reactors' biggest advantage is their simple design. The gas streams through a layer of small catalyst pellets which leads to a varying heat development and thus, to hot spots. Furthermore, fluctuating loads are hard to handle for this reactor type. Fluidized-bed reactors are based on the same principle, but the pellets are much smaller and the fluid enables a homogeneous heat transport. However, their disadvantage is that the increased mechanical stress on the catalyst pellets can lead to their destruction. Different loads can be problematic for this technology as well, thus the two-phase catalytic or chemical methanation is not suited for use with renewable energy sources.

In a bubble column reactor, the gas streams through a liquid oil with solid catalyst particles which results in a bubble column. The heat conducting oil is a good heat dissipater and the process is very flexible in the amount of gas flow.

11.4.3.2 *Biological methanation*

Biological methanation is based on the same reactions as the catalytic driven variant, but in this process methanogenic microorganisms serve as a biocatalyst. In a conventional biogas plant, the hydrolysis of an organic substrate to simple monomers begins the process chain. Subsequently, the monomers (monosaccharide, amino acids and fatty acids) are converted into CO_2, CH_3COOR and H_2. The methane is produced by aceticlastic methanogenesis, the depletion of acetate and hydrogenotrophic methanogenesis. Biological methanation only proceeds between a temperature of 20 to 70°C and an anaerobic environment. Nonetheless the technical implementation is not trivial and is still an issue to be worked at. The efficiency depends on the type of microorganism, cell concentration, pressure, reactor concept, pH-value and various other factors. A key parameter for the evaluation of a reactor is the methane formation rate (MFR).

$$MFR = \frac{F_{V,CH_4,out} - F_{V,H_2/CO_2,in}}{V_R}, \qquad (11.10)$$

where F is the volumetric flow rate and V_R is the reactor volume.

Two main process concepts have been the focus for academic as well as industrial research. On one hand, the methanation is using an in situ digester, while on the other, the methanation is in a seperate reactor [Goetz et al. (2016)].

11.4.3.3 *Plasma-based methanation*

Plasma is a partially ionized gas, consisting of a huge number of different species which can all interact with each other, resulting in a highly reactive mixture. The advantage of plasma technology for CO_2 conversion is based on the presence of highly energetic electrons, which are able to activate the gas by electron impact ionization, excitation and dissociation. Thus, the gaseous CO_2 itself does not have to be heated as a whole, but can remain near room temperature. Due to this process, even strongly endothermic reactions, like the decomposition of CO_2, can occur with low energy consumption and without extreme reaction conditions. The sub-product CO then reacts with externally generated H_2 and freely forms methane.

Several types of plasma have been applied for this purpose; the most common ones are dielectric barrier discharges [Siemens (1857)], glow discharge plasma [Speckmann et al. (2017)], microwave plasma [Kosak and Bogaerts (2014)] and gliding arc discharges [Nunnally et al. (2011)]. All plasma technologies have a low energy efficiency of only 10–20% and this is too low for industrial implementation. On the other hand, this type of methanation is still in progress and enables the synthesis of hydrogen from electric energy, carbon dioxide and water, without any additional requirements. In an energy mix with an increased amount of renewable energy sources, these technologies have a promising future.

11.5 Power-to-Liquid

The change from fossil energies towards renewable energy sources is still at the infancy phase, with regard to the transport sector. Only 5–10% of the fuels are generated from sustainable sources, and most of them are used in electric railway transportation and biofuels. Therefore, the dependence on mineral oil (90%) is quite high and leads to immense emissions of climate gases.

There are different strategies to lower the emissions from this sector. Besides electric vehicles, there is the possibility of using renewable fuel from PtL processes. Cars, trucks, ships and airplanes are already equipped with conventional fuel tanks and only need to be slightly modified in order to use these newly generated fuels. Thus, this makes synthetic fuels a possible solution to sustain the transport sector in the future, or act as a short-term solution until electric vehicles are more advanced and cost-efficient. Predictions for the transport sector's future energy demands show a decline in total energy consumption as well as the increased use of synthetic biofuels, as shown in Fig. 11.6 [Nitsch et al. (2011)].

Furthermore, gas stations might then be using their standard fuel storage tanks and would not need to invest into hydrogen tanks or charging infrastructure for electric cars.

11.5.1 *Technological overview*

The PtL process incorporates the energy chain from electricity conversion to mostly hydrogen, following the synthesis into a fuel. In most PtL systems the electricity is fed to some kind of electrolysis, the major ones being AEL,

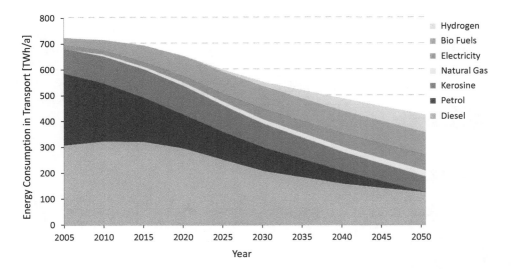

Fig. 11.6 Estimated energy consumption in the transport sector separated into energy carriers.

PEMEL and HTEL, as described in Sec. 11.4.2. The generated hydrogen, together with carbon dioxide, is then supplied to a water gas shift reactor or a methanol synthesis, depending on whether the system is Fischer–Tropsch or methanol-based. A water gas shift reaction of the generated hydrogen with carbon dioxide is required for a subsequent Fischer–Tropsch synthesis. Another possibility is that of the hydrogen and carbon dioxide being directly fed into a methanol synthesis. Both processes are displayed in Fig. 11.7.

The key steps in the first conversion scheme are the water-gas shift reaction (see Eq. 11.11)

$$CO + H_2O \longleftrightarrow CO_2 + H_2, \qquad \Delta H_R = -41.2 \, kJ \qquad (11.11)$$

as well as the Fischer–Tropsch synthesis (Eq. 11.12).

$$(2n+1)H_2 + nCO_2 \longleftrightarrow C_nH_{2n+2} + nH_2O. \qquad (11.12)$$

The entire process involves different chemical reactions that result in a variety of hydrocarbons; the more useful reactions produce alkanes, where n is typically between 10 and 20. These alkanes are often straight chains that are suitable as diesel fuel. Other possible reaction products are alcohols or a wide range of oxygenated hydrocarbons [Kaneko et al. (2001)].

The methanol synthesis is a process that is based on the heterogeneous catalysis of CO_2. A conventional methanol synthesis requires a high performance catalyst, which has to be highly active but stable for a long period of

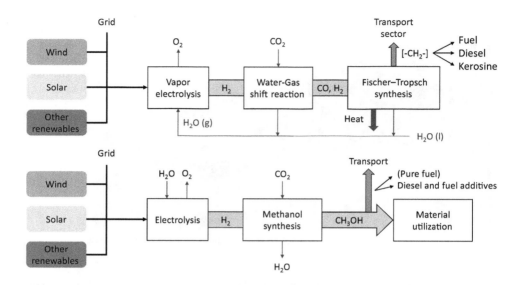

Fig. 11.7 Conversion chains of Power-to-Liquid processes.

time, as well as selective for the methanol generation. The used catalysts vary a lot, although most modern Cu/ZnO multicomponent catalysts contain two or three metal oxides (e.g., $Cu/ZnO/ZrO_2/Al_2O_3$) [Saito et al. (1997)].

The feed for the methanol synthesis is a $CO/CO_2/H_2$ mixture, with CO as a reducing agent, which leads to the following reaction steps shown in Eqs. (11.13) to (11.15).

$$CO_2 + 3H_2 \longleftrightarrow CH_3OH + H_2O, \tag{11.13}$$

$$CO_2 + 2H_2 \longleftrightarrow CH_3OH + O_{(a)}, \tag{11.14}$$

$$CO + O_{(a)} \longleftrightarrow CO_2. \tag{11.15}$$

(a) stands for activated. Therefore, the net reaction is Eq. (11.16)

$$CO + 2H_2 \longleftrightarrow CH_3OH. \tag{11.16}$$

Using several commercial copper-based catalysts, it was proven that no methanol could be produced using syngas without CO_2, and from which all water was withdrawn. In addition, isotopic labeling proved that CO_2 is the source of carbon in the methanol [Spath and Dayton (2003)]. At high CO_2 concentrations, it reduces the catalyst activity by inhibiting methanol synthesis. Thus, the conversion of CO and CO_2 to methanol is limited by the pertinent chemical equilibrium as a function of the temperature.

Dimethyl ether (DME) can have similar uses as a flexible fuel in a wide variety of fields and is produced by a similar process chain. It is a clean energy source as it generates no sulfur oxide or soot during combustion. Therefore, its environmental impact is low. Existing technologies produce DME through the dehydration of methanol, as shown in Eq. (11.17).

$$2CH_3OH \longleftrightarrow CH_3OCH_3 + H_2O. \qquad (11.17)$$

DME is easy to handle and could be a domestic-sector fuel substitute for liquid propane gas or diesel. However, the scale of operating plants nowadays is rather small and scaling-up is an issue. Most projects plan on using natural gas as a feedstock because of its low initial investment costs. However, most of them are expected to switch to coal in the future, because it has larger reserves than natural gas. Using fossil resources leads to additional CO_2 emissions and is rather suboptimal. When connected with methanation systems, the coal-based process would be able to provide purified CO_2 without any additional equipment [Bertau et al. (2014)].

11.5.2 Carbon sources

There are various CO_2 sources that can either be extracted from the atmosphere or be gained from biogeneous or fossil conversion processes. The combustion of fossil energy carriers results in the generation of CO_2. It is possible to extract the CO_2 directly from the exhaust gas and use it for further conversion. However, the synthesized gas from the extracted exhaust gas is not regarded as biogas by current European law. Furthermore, is it necessary to buy CO_2-certificates for the combustion of this type of SNG, which results in additional costs. In contrast, with regard to the SNG from exhaust gases, does the CO_2 from biogeneous sources, together with hydrogen generated from renewable energy via electrolysis, count as biogas? Therefore, will it be common to use CO_2 from the fermentation of plants, where produced gas consists of about 70% methane and 30% CO_2? This CO_2 content has to be extracted in order to be able to feed the remaining methane gas to the local gas grid. There is no commercial use for the remaining, highly concentrated CO_2 as of today. Thus, it can be utilized for methanation as soon as all possible impurities are removed. In 2011, 80 out of the 7215 biogas plants in Germany were biomethanol plants that provided 275 Mio. Nm^3 biomethane to the gas grid. This led to an extracted amount of 160 Mio. Nm^3 of CO_2 [(Ulleberg (7)]. The amount of CO_2 from these 80 plants enables a potential of 3.3 TWh of electricity to be converted in SNG without costs for the CO_2.

There are no dedicated specific investment costs for methanation units right now. According to Sterner (2014), the investment costs for a demonstration plant (5–10 MW_{el}), without a re-electrification unit, are about 2000 $Euro/kW_{el}$. Electrolyzers cost in the range of about 1000 Euro/kWh, thus, leaving the same amount for the methanation system and the H_2 backup storage. The investment costs for the electrolyzers as well as the methanation process are predicted to decline with larger systems (20–200 MW_{el}) and more research in this area.

11.6 Power-to-Solid

The PtS technology aims for energy storage via a solid medium as well as a subsequently reconversion. It can enable a hydrogen or power-on-demand system. Such a technology would not require a high-pressure gas storage tank, it would simply need as much hydrogen as a fuel cell currently requires. There are other different approaches in this field of research, varying in the choice of solid medium and the kind of energy they provide. Most PtS systems can be arranged in one of the three characteristic processes shown in Fig. 11.8.

Firstly, the energy chain from electricity to the charging of a metal-based battery, commonly metal-air batteries. These directly provide electricity

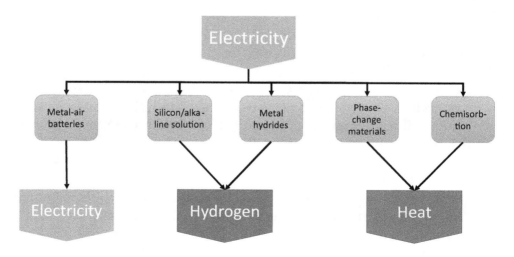

Fig. 11.8 Conversion chains of Power-to-Solid processes.

when discharged and therefore are key components for several applications. The major application for metal-air batteries nowadays is the use of zinc-air cells for hearing aid devices, for example. These one-way systems are only usable for a set duration until their capacity is exhausted and can only supply a voltage of about 1.4 V. Every year, billions of zinc-air cells are produced by companies such as VARTA. Another example are aluminum-air batteries provided by Phinergy. The aluminum-air technology uses naturally occurring oxygen to fill its cathode. Aluminum-air batteries' discharge turns the metal into aluminum hydroxide, which can then be recycled to make new batteries. This makes it far lighter than liquid-filled lithium-ion batteries to give the car a far greater range [Levitt (2017)]. Though one of its disadvantages is that the cell requires replacing every few months, at least its components are recyclable. This generally shows the challenges that face such systems. As soon as the discharge process is started, it cannot be stopped easily.

Secondly, the electricity can be stored via the utilization of hydrogen. One possible conversion process is that of an electrolyzer linked to a metal hydride storage. Metal hydrides, such as MgH_2, $NaAlH_4$, $LiAlH_4$, LiH, $LaNi_5H_6$, $TiFeH_2$ and palladium hydride, with varying degrees of efficiency, can be used as a reversible storage medium for hydrogen [Zaluska et al. (2001)]. The H_2 bounds strongly with most metal hydrides that require temperatures of around 390 K in order to release the H_2 content. Activators or metal alloys are investigated for the reduction of the required temperature. These systems have good energy density by volume, although their energy density by weight is worse than the leading hydrocarbon fuels. A different approach uses electric energy to produce silicon crystals that react with alkaline solutions to hydrogen. This method enables the above-mentioned H_2 generation on demand. Fuel cells can be used for a reconversion to electricity, or an engine is supplied for a direct use, e.g., mobile applications.

Thirdly, the storage process can be based on heat energy with the help of phase change materials (PCM). These substances are capable of storing and releasing large amounts of energy while melting or solidifying at certain temperatures. Heat is absorbed or released when the material changes from solid to liquid, and vice versa. Therefore, PCMs are classified as latent heat storage units. Most organic materials like paraffin or carbohydrates are chemically stable as well as recyclable, but only have low thermal conductivities in their solid states. Furthermore, the volumetric latent heat storage capacity is low. The inorganic PCMs are mostly salt hydrates which are inexpensive and widely available. They are non-flammable and have a high thermal conductivity. Their disadvantage is their high change in volume and they often become inoperative after repeated cycling.

11.7 Basic Gas Management Systems

Gas management systems (GMS) can be compared to conventional battery management systems with the tasks of protecting the battery module from operating outside its safe-operating-area, monitoring its state, calculating secondary data, reporting that data, controlling its environment, and authenticating and balancing it [Rahimini-Eichi *et al.* (2013)]. The main tasks of a GMS are the regulation of gas flows, controlling pressures, checking safety settings and communication between the user and the systems regulation electronics. Such a system has to be adopted for special applications ranging from fuel cell systems to motor engines and stationary gas power plants.

The functionality of a GMS will be highlighted with the example of a "silicol" process for mobile applications. Silicon granulate, in conjunction with a sodium hydroxide solution ($NaOH$), produces H_2, as shown in Eq. (11.17).

$$Si + 2NaOH + H_2O \longleftrightarrow Na_2SiO_3 + 2H_2. \qquad (11.18)$$

This reaction consumes the alkaline solution and forms the water-soluble sodium silicate. Stationary "silicol" systems have already been exemplified. The setting for the mobile application consists of storage tanks for the sodium solution, a reaction chamber and a fuel cell stack, as displayed in Fig. 11.9.

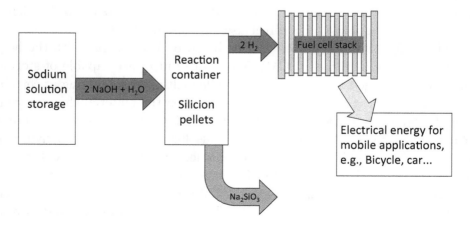

Fig. 11.9 Hydrogen generation from solid materials for mobile applications.

The sodium solution has to be pumped into the reaction container containing the silicon pellets, in order to start the H_2 generation. An increasing pressure in the reaction chamber leads the gas through a controlled valve into the fuel cell. The fuel cell (PEMFC) has an integrated reducing adapter which can regulate the maximum gas input. An electric current with a certain voltage is further provided for the electric drive of the vehicle.

There are several connections between the described set-up and the GMS. The controlling unit, in this case a micro controller, is directly attached to the two pressure sensors, the temperature sensor, the level measurement of the sodium solution tank, a DC-DC-converter, the fuel cell stack, as well as the fast switching valve (cf. Fig. 11.10, dashed lines). The converter unit increases the voltage level required for the pump and valve 1 (cf. Fig. 11.10, see dashed lines at (a)). A process cycle of this system works as follows: sodium solution is pumped into the connection between the tank and the reaction chamber (valve 1 is closed). The check valves 1 and 2 prevent the backwards flow. Afterwards, valve 1 is opened and the pump turned off. The high pressure $NaOH$ is released into the reaction container. The hydrogen forming reaction takes place and increases the pressure in the container. After a certain pressure is detected by the pressure sensor 1, valve 2 is opened for a brief moment. If the pressure level in the reaction chamber is below a certain value, more sodium solution is injected. Pressure sensor 2 detects the pressure at the fuel cell and gives a feedback signal to the controller. Thus, a two-step control is realized, which ensures a constant gas flow for the fuel cell and enough power for the electric engine. Furthermore, are the filling level of the sodium solution tank and the temperature in the reaction tank are detected. Therefore, feedback for the user is possible, and if the system runs low on "fuel" or if the heat rises to a critical limit, then the system has to be shut down. The integration of the GMS into the mobile application is displayed in Fig. 11.10.

11.8 Sustainable Energy Chains — Closing Remarks

This chapter shows how sustainable use of energy can lead to increased efficiency of industrial supply chains, improved financial profitability and a "greener" future. This chapter provides comprehensive coverage of current practices and possible future developments for the evolution of sustainable supply chains, energy storages, as well as energy consumption.

Renewable energy technology is a fast growing market with promising financial returns and substantial environmental gains. The cross-sectoral concept of PtX highlights possible integration of "green" energy in a wide

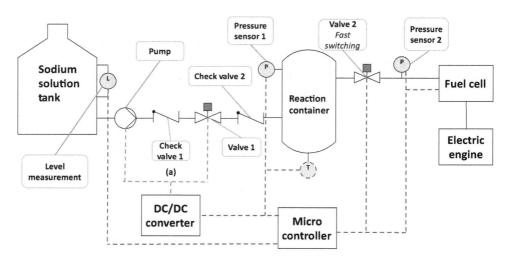

Fig. 11.10 Gas management system for hydrogen generation from solid materials for mobile applications.

field of applications. Very important is the fact that renewable energy generation is directly connected with energy storage technologies and the conversion of it into different energy carriers. In order to establish sustainability for the various energy needs of a society, the whole process from generation, together with transportation, the conversion and finally the consumption has to be taken into account. Establishing a transport sector based on electric vehicles alone does not provide a CO_2 neutral future. It is necessary that the electricity is generated from renewable energy sources such as photovoltaics or wind turbines. Additionally, can PtX systems even decrease the amount of CO_2 in the atmosphere and combine it with H_2 into value-added fuels and chemicals? Whether batteries are used today in electric vehicles or biomethanol as a sustainable fuel, and if men hope to eliminate their dependence on fossil fuels in the future, then energy storages of any kind need to be found. Not one technology can achieve that goal alone, but a mix of different technologies with special benefits for certain applications can.

Bibliography

Barbir, F. (2005). PEM electrolysis for production of hydrogen from renewable energy sources, *Sol. Energy* **78**, pp. 661–669.
(eds.) Bertau, M., Offermanns, H., Plass, L., Schmidt, F., and Wernicke, H. J. (2014). *Methanol. The Basic Chemical and Energy Feedstock of the Future* (Springer-Verlag, Heidelberg).

Bossel, U. (2003). Well-to-wheel studies, heating values, and the energy conservation principle, European Fuel Cell Forum, Switzerland.

Entsoe (2017). Network Codes, http://networkcodes.entsoe.eu/category/introducing-network-codes/.

Gardiner, M. (2012). Fuel cells and hydrogen enabling large scale renewable energy, IPHE, US Department of Energy Fuel Cell Technologies Program, Sevilla.

Goetz, M., Lefebvre, J., Moers, F., McDaniel Koch, A., Graf, F., Bajohr, S., Reimert, R., and Kolb, T. (2016). Renewable Power-to-Gas: A technological and economic review, *Renew. Energy* **85**, pp. 1371–1390.

Kaneko, T., Derbyshire, F., Makino, E., Tamura, M., and Gray, D. (2001). *Coal Liquefaction, Ullmanns Encyclopedia of Industrial Chemistry*, doi:10.1002/14356007.a07_197, http://onlinelibrary.wiley.com/doi/10.1002/14356007.a07_197/otherversions.

Kosak, T. and Bogaerts, A. (2014). Evaluation of the energy efficiency of CO_2 conversion in microwave discharges using a reaction kinetics model, *Plasma Sources Sci. Technol.*, **24**, 1, pp. 17–34, http://iopscience.iop.org/article/10.1088/0963-0252/24/1/015024/meta.

Kreuter, W. and Hofmann, H. (1998). Electrolysis: The important energy transformer in a world of sustainable energy, *Int. J. Hydrogen Energy* **23**, pp. 661–666.

Ladomenou, K., Natali, M., Iengo, E., Charalampidis, G., Scandola, F., and Coutsolelos, A. G. (2015). Photochemical hydrogen generation with porphyrin-based systems, *Coord. Chem. Rev.* **304–305**, pp. 38–54, doi: https://doi.org/10.1016/j.ccr.2014.10.001, https://www.sciencedirect.com/science/article/pii/S0010854514002732.

Larminie, J. and Dicks, A. (2003). *Fuel Cell Systems Explained*, 2nd Edition (Wiley, UK).

Levitt, J. (2017). Israels Phinergy tests 1100-mile range electric car; aluminum-air battery system, https://www.algemeiner.com/2014/06/17.

Marini, S., Salvi, P., Nelli, P., Pesenti, R., Villa, M., Berrettoni, M., Zangari, G., and Kiros, Y. (2012). *Electrochim. Acta* **82**, pp. 384–391.

Nitsch, J., Pregger, T., Scholz, Y., Naegler, T., Sterner, M., Gerhardt, N., von Oehsen, A., Pape, C., Saint-Drenan, Y.-M. and Wenzel, B. (2011). Leitstudie 2010 (in German), Projektbericht im Auftrag des Bundesministeriums fr Umwelt, Naturschutz und Reaktorsicherheit (BMU), BMU, Germany, http://www.ifne.de/download/Leitstudie_2010.PDF.

Nunnally, T., Gutsol, K., Rabinovich, A., Fridman, A., Gutsol, A., and Kemoun, A. (2011). Dissociation of CO_2 in a low current gliding arc plasmatron, *J. Phys. D: Appl. Phys.* **44**, 27, 4009, doi: https://doi.org/10.1088/0022-3727/44/27/274009.

(eds.) Pachauri, R. K. and Reisinger, A. (2007). IPCC, 2007: Climate change 2007: Synthesis report, Report of the Intergovernmental Panel on Climate Change, IPCC, Geneva, Switzerland, 104 pp.

Rahimini-Eichi, H., Ojha, U., and Baronti, F. (2013). Battery management system: An overview of its application in the smart grid and electric vehicles, *IEEE Ind. Electron. Mag.* **7**, pp. 4–16.

Saito, M., Takeuchi, M., Watanabe, T., Toyir, J., Luo, S., and Wu, J. (1997). Methanol synthesis from CO_2 and H_2 over a CuZnO-based multicomponent catalyst, *Energy Convers. Manage.* **38**, pp. 403–408.

Siemens, W. (1857). Ozone production in an atmospheric-pressure dielectric barrier discharge, *Poggendorff's Ann. Phys. Chem.*, **102**, p. 66.

Spath, P. L. and Dayton, D. C. (2003). Preliminary screening technical and economic assessment of synthesis gas to fuels and chemicals with emphasis on the potential for biomass derived syngas, National Renewable Energy Lab Golden Co., Golden, US, http://dx.doi.org/10.2172/1216404.

Specht, M., Zuberbuhler, U., Baumgart, F., Feigl, B., Frick, V., and Stuermer, B. (2009). Storing renewable energy in the natural gas grid, *FVEE-AEE*, pp. 69–78, http://www.fvee.de/fileadmin/publikationen/Themenhefte/th2009-1/th2009-1_05_06.pdf.

Speckmann, F.-W., Mueller, D., Koehler, J., and Birke, K. P. (2017). Low pressure glow-discharge methanation with an ancillary oxygen ion conductor, *J. Utilization*, **19**, pp. 130–136.

Sterner, M. and Stadler I. (2014). *Energiespeicher, Bedarf Technologien Integration*, 1st Edition (Springer Vieweg, Germany).

Ulleberg, Ø. (2003). Modeling of advanced alkaline electrolyzers: A system simulation approach, *Int. J. Hydrogen Energy* **18**, pp. 21–33.

Vincent, C. A. and Scrosati, B. (2014). *Modern Batteries*, 2nd Edition (Butterworth-Heinemann, UK).

Wietschel, M., Ullrich, S., Markewitz, P., Schulte, F., and Genoese, F. (2015). *Energietechnologien der Zukunft*, 1st Edition (Springer Vieweg, Germany).

Zaluska, A., Zaluski, L., and Ström-Olsen, J. (2001). Structure, catalysis and atomic reactions on the nano-scale: A systematic approach to metal hydrides for hydrogen storage, *Appl. Phys. A.* **72**, p. 157–165, doi:https://doi.org/10.1007/s003390100.

Zapf, M. (2017). *Stromspeicher und Power-to-Gas im deutschen Energiesystem, Rahmenbedingungen, Bedarf und Einsatzmglichkeiten*, 1st Edition (Springer Vieweg).

Epilogue

The future of Lithium-ion cell technology

At first glance, energy density appears to be the main driver for the next generations of batteries, which makes sense and is intuitively right. Therefore, the focus is still often only on gravimetric energy density (Wh/kg), though volumetric energy density (Wh/l) is increasingly becoming the decisive target, especially regarding automotive applications. An electric vehicle has to appear attractive, let's even say sexy, and it has to provide enough space for its passengers and their luggage. This is only possible by employing a battery with sufficient volumetric energy density. A further but widely neglected aspect is the correlation between energy density and costs, which is in €/kWh or ct/Wh. Assuming that energy density can be continuously enhanced without employing a completely new electrochemical system, the cells and battery costs will consequently decrease. This is a major insight in battery technology and is the essential reason to optimize energy densities.

Sophisticated argumentation is crucial

However, the pure focus on energy density is one of the biggest sources of disharmony between intensive lab-scale research on future battery-cell generations and industrial-related development of all-day reliable electrical energy storage systems (otherwise known as batteries). Despite this, energy still plays a pivotal role in the direction of future research on electrochemical systems and how batteries should be seen nowadays. In this regard, even the three additional major criteria of importance — power density, lifetime expectation and safety — need not to be considered in-depth.

Now, we should clarify the meanings of "cell" and "battery". A cell is the smallest electrochemical unit. A battery describes the interconnection of cells. An electrical energy storage system is a battery with additional features such as electrical and thermal battery management, housing, cooling and optionally heating devices, and passive protection elements like switches and fuses. Therefore, it is of great importance to distinctly distinguish between a cell and a battery. As an example, although there is no large-scale cell production unit in Europe, there are still leading companies that produce batteries in Germany.

Despite tremendous know-how and intensive research activities devoted to the battery, the cell still represents about 70 per cent of the cost distribution in a battery. In future, this value will increase rather than decrease. Apart from the quality aspect of delivered cells, this is one of the most prominent arguments to achieve a national own-cell production. However, it has to be carefully considered that in the cell itself, about 70 per cent of the value chain concentrates on raw materials and their refinement to cell materials, especially the active (contributing to the electrochemical reaction) materials in the electrodes, the liquid electrolyte and the separator, and the metallic current collectors. To summarize, we have to intensively care about mining companies and materials refinement to battery grades; access to cobalt and nickel is already actually of extreme importance.

With this background in mind, the huge efforts to develop Li-sulphur and Li-air cells appear in another light. Sulphur is a waste product from the industry and is thus extremely low-cost, as compared to the commonly used active cobalt-, nickel- and manganese-based cathode materials in Li-ion cells. Even oxygen in the air is free of charge! But here, we find the differences between a purely materials and purely energy density driven points of view.

Often, only so-called theoretical energy densities are considered. Though commercial cylindrical Li-ion cells already exceed energy densities of 250 Wh/kg, one can surprisingly find values of 11140 Wh/kg for Li-air (also known as Li-oxygen) and 2600 Wh/kg for Li-sulphur, respectively. However, values coming from chemical equations solely based on active materials are mixed up with the values of complete cells, which are applicable in batteries. But in reality, a technology must be ranked by values that are achievable on cell level (as a commercial product). As an example of the "magic" energy loss from paper to battery, we look at a lead acid battery (see Fig. 2).

With this approach we see another reality. At first it should be emphasized that only 3–4 weight per cent of metallic lithium are present in a commonly known modern Li-ion cell. The rest is simply spoken as "the big

Fig. 1 Sulphur would theoretically be one of the most attractive cathode materials in Lithium-based cells. As an example, the figure depicts naturally occurring sulphur in Sicily.

show". The theoretical energy density of pure lithium metal would be around 16000 Wh/kg, which even exceeds the energy densities of gasoline or diesel (about 10000 to 12000 Wh/kg). Only hydrogen with 33000 Wh/kg is "top of the pops" and this is the main reason for the attractiveness of fuel cells

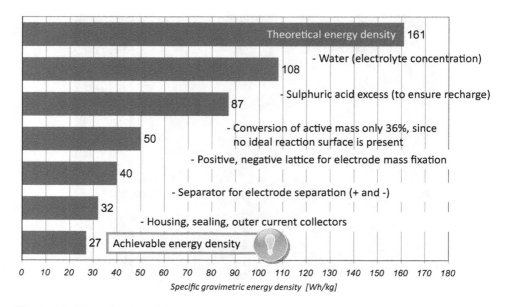

Fig. 2 Taking a lead acid battery as an example, the loss of energy density starting from a theoretical view and based on the active materials chemical reaction equation towards the feasible battery.

in traction applications, especially for heavy duty, long distance and heavy vehicles or trains. Three weight per cent lithium theoretically provide about 500–640 Wh/kg, in case of actual active so-called intercalation or insertion materials that are graphite on the negative, and a lithium transition metal oxide on the positive electrode in a commercial high energy density Li-ion cell. The name Li-ion results from this rocking-chair principle that swings Li-ions between the electrodes like water in a sponge. These cells are different from those applying pure Li-metal (such as Li-sulphur and Li-air) and this must be carefully distinguished. A particular property of Li-ion cells is the fact that the volumetric energy density counts about twice as compared to gravimetric energy density (see Fig. 3), which is a decisive property if electric vehicles are focused upon.

But, are Li-sulphur and Li-air realistic alternatives?

Exactly, this does not hold true for the often-discussed "future candidates" Li-sulphur and Li-air, because the ratio between volumetric and gravimetric energy density is only about 1:1. Taking into account that sulphur is electrochemically nearly inert, thus considerable amounts of conducting agents (carbon) and electrolyte are required to sufficiently activate the

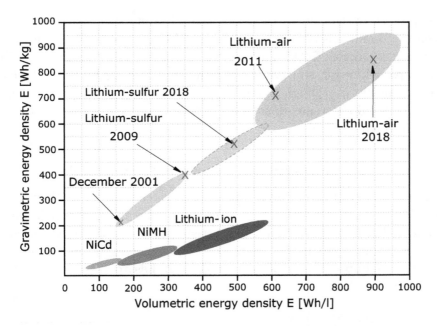

Fig. 3 Volumetric and gravimetric energy densities of different cell types.

sulphur we have in the situation that feasible Li-sulphur cells can still show an advantage over Li-ion cells, regarding the gravimetric energy density (300–400 Wh/kg). However, as far as volumetric energy density is concerned, commercial Li-ion cells showing 600 Wh/l (or more) already have distinctly succeeded over Li-sulphur cells. As for automotive applications, the volumetric energy density has clearly become the actual focus since we will never see any Li-sulphur batteries powering electric vehicles. At this point, it is not even necessary to compare the power capability and cyclability of Li-sulphur and Li-ion cells. One can find the following picture for Li-air cells: since oxygen must be stored upon every cell discharge, there is a tremendous loss of energy density from 11140 Wh/kg to 3391 Wh/kg, assuming the reaction to the peroxide Li_2O_2. Taking also into consideration that Li_2O_2 is electrochemically nearly inert, but cannot be placed into a conductive matrix such as in the case of Li-sulphur (because the reaction with oxygen needs a large surface and interaction with air, the cell has to be semi-open), one has to apply thin layers (about 50 nm) and catalytic materials. Thus achievable energy densities of lab-scale cells are even less attractive than those of commercial cylindrical Li-ion cells, in which the electrode bulk can be used and compressed thick electrodes (of up to 200 µm) can be applied. Additionally, typical catalytic materials are gold, platinum or other noble metals. Thus, all challenges of an advanced battery cell and a fuel cell are concentrated and this is the reason why any breakthrough — if at all — will not take place before 2030. The big question of whether Li-air will still be competitive will arise, since other systems have already been developed for ten years (or more), though Li-air is on paper, a fantastic advertisement for a system providing tremendous energy density (in the charged state), and oxygen in the air is free of charge.

Attractive electromobility has become battery feasible

The core message of this approach is that Li-ion cells, as far as their already achievable energy densities in commercial cells are concerned, already dominate plenty of so-called future systems. In principle, only discussions about solid-state systems still seem to be both necessary and essential, as seen from their prominent place in Fig. 4. It has to be emphasized that

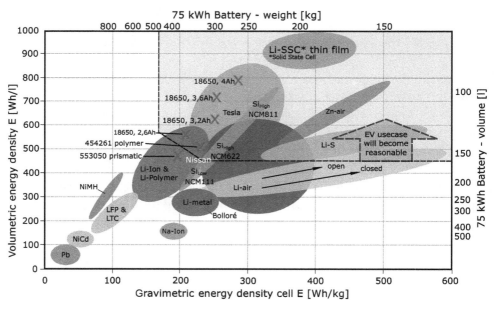

Fig. 4 This figure depicts battery feasible electromobility energy densities that have already been achieved by commercial systems and compares them with potential future approaches. At the same time, cell and system data are brought into a useful correlation. The crosses show that cylindrical cells are already present (Tesla approach), which enables useful electromobility (upper right area in the diagram). At the same time it can be shown that for energy density achievement on cell level, Li-sulphur, Li-air and Zn-air cannot really compete with Li-ion. Solid-state-based lithium cells (SSC) try to overcome this, ever since Toyota announced its intention to develop SSCs with a focus on thin films.

many solid-state systems make use of the big advantage that Li metal anodes can be employed. The extraordinary specific capacity of metallic lithium of 3380 mAh/g, together with the fact that lithium is the most electronegative element, makes the energy density (which is voltage times specific capacity) not the solid ion conductor. This is also the simple reason for the high energy densities of Li-sulphur and Li-air, which are only achievable in combination with metallic lithium anodes. Solid-state systems could suppress the well-known side reactions of systems with liquid organic electrolytes upon cycling and thus can effectively make use of the distinct energy density push coming from the metallic lithium electrode. This would also be a tremendous push versus conventional lab-based Li-sulphur and Li-air cells that still mainly rely on liquid organic electrolytes and suffer from poor rechargeability and minor safety issues. But metallic lithium is the development driver, not the solid ion conductor. Thus, if liquid or gel-type electrolytes were to be successfully employed together with a metallic lithium electrode, the solid-state approach would become obsolete.

Solid ion conductors — vision or nightmare?

Successfully operating cells with solid ion conductors have been known for decades. During the 1970s, the Na-S and the ZEBRA-battery (Na-NiCl$_2$) have been developed in well operating systems, although temperatures of around 300°C are necessary. Some years ago, Bolloré, a French company, presented a solid-state system based on metallic lithium as a negative electrode (anode) and lithium iron phosphate (LFP) as a positive electrode (cathode), employing a PEO (polyethyleneoxide)-based so-called solid-state ion conductor that has even been commercialized. LFP operates only about 3.2–3.3 V as opposed to lithium; this is the reason why PEO stays stable in such a system. A start-up SEEO from California took up this idea and worked out options to make so-called standard cathodes (NCM, Lithium-nickel-cobalt-manganese-oxide) in Li-ion cells accessible for a PEO-based system. Two different approaches were used, either by coating of the NCM and/or by modification of the PEO's border on the cathode side. Partial successes from these approaches may have been the reason for SEEO being acquired by Bosch. Unfortunately, this arrangement has since been terminated.

Today, the huge challenge is to make these systems capable for operation at temperatures lower than 60°C. A case in point is that a compromise of using battery heating systems profiting from high internal impedance at lower temperatures ended up a failure because the cell impedance was too high at 20°C, or even 40°C, to operate such a heating system successfully — the system couldn't be self-heated by the battery cells anymore. Since their high impedance at 20°C, or even 40°C, blocks sufficient heating currents, an advanced external heating system becomes necessary. Additionally the temperature (of about at least 60°C) has to be carefully controlled and preserved under operation conditions, in order not to become too low until the battery is unable to satisfy the power requirements anymore. This, together with the challenge of incorporating metallic lithium into the cells by a roll-to-roll production device and the all-time "evil" of potential lithium dendrite growth (growth of fine metallic needles towards the cathode causing short circuits and even severe safety issues) when using a polymer solid electrolyte, resulted in the system only having a degree of maturity which would have required strong long term efforts. This may have been the technical reason for Bosch to exit its engagement in SEEO.

Generally the breakthrough of solid ion conductors is closely connected to the feasible operation temperature and the successful suppression of dendrite growth. In addition, ceramic-based solid ion conductors face huge challenges regarding weight and material costs (e.g., lanthanum and zirconium are main parts of them). It is a fairy tale come true that these solid ion conductors are, in principle, easy to handle. Since the development target of, for example, lithium-lanthanum-zirconium-oxide (LLZO) is its stability with metallic lithium, these compounds are overfed with lithium, which means that there is a strong chemical force that will release lithium. This leads to the formation of strongly aggressive and insulating lithiumhydroxide in contact with all-time present traces of water and in contact with air to insulating lithiumcarbonate, which is a nightmare since solid Li-ion conductors materials are usually powders and strongly dependent on particle-to-particle contacts in a cell. In addition, lithiumhydroxide can trigger undesired side reactions in the cell. However, subsequent research activities in the field of solid ion conductors could raise an interesting alternative to conventional Li-ion cells if attractive energy densities were to rule this system; instead of a sophisticated cooling system, a simple, cost-effective heating and insulating concept on the battery system level, plus distinct improvements of high temperature stability (>60°C), would result in safety and cyclability being achieved.

But solid ion conductors are not miracles or the essence of the master plan. For sure, they are not non-toxic or inert all the time, hence the required raw materials have to be carefully watched in terms of costs and abundancy. This also means that in the event of solid ion conductors undergoing undesired electrochemical reactions with the electrodes can be present. And due to the reduced electrolyte/electrode contact area, there is a huge sensitivity to any volume changes (which is drastic in the case of cycling a metallic lithium anode) with respect to losing contact. All in all, tremendous challenges in developing all solid-state cells remain.

Why carry out Li-ion cell production in Germany?

Considering the initially delivered arguments that a Li-ion battery cell is in principle only a refinement of raw materials that holds a remarkable part of the value chain of the cell, why carry out cell production in Germany, which is in Europe? Wouldn't a buy-in concept of Li-ion cells from countries (e.g., China) that have the best access to these raw materials be a better option? There are nonetheless interesting arguments for this:

The cell production must take place near battery production and devices where these batteries are implemented, e.g., vehicles and stationary applications

Li-ion cells have to be produced near their fields of application because due to strict transport rules, costs, long transport distances (time!) and potential damage of cells (e.g., in case of overseas transport combined with high temperatures in non-conditioned containers), severe disadvantages arise. Therefore, the well-known big players in the cell business have already built up cell factories in Europe.

Avoidable costs, being and staying dependent and loss of know-how

Purchasing cells always brings up a cost disadvantage since additional cost for the cell manufacturer must be factored into the overall battery cost structure. And if the battery becomes similarly meaningful for the diversification of electric vehicles as the combustion engine earlier and still nowadays for conventional vehicles, then an individual electrochemical cell design is strongly required for the OEMs. These design principles should not simply be given away to an outside cell manufacturer who will produce cells using an OEM's proprietary knowledge. In an even worse case scenario, the OEM becomes dependent of the performant cells of a certain cell manufacturer. However, it is remarkable that in Germany — and the same may hold true for other European countries — there is only a gap regarding the mass production of cells but not in the know-how gained from intensive cell research, cell design and machine manufacturing.

A further decisive quality push is strongly recommended and necessary for Li-ion cells

In the wake of increasing demand for Li-ion cells and the fact that modern day batteries consist of many single cells (Tesla batteries have 7000–10000 cylindrical cells per battery. Stationary batteries in the regime of 1 MWh require even more single cells.), the quality improvement of any cell during its production process becomes of tremendous strategic importance. The reason relates to statistical failure: x cells per fixed produced number of cells, meaning though this percentage remains constant, the number of cells behind is increasing drastically, and since the produced cells increase dramatically, so does the failures and the number of bad cells. This is especially the case in large numbers of serial connections, where the batteries will be strongly affected.

Such an increase in quality can be achieved and performed by means of digitalization and industry 4.0-compatible approaches. By this, I mean that a tremendous amount of cell data can be collected, a path back (e.g., where failures have been caused) can be identified, quality issues can be detected earlier and that cell production will one day become self-learning and self-optimizing. This will mix up the cards again and this will provide Europe, especially Germany, with a unique chance of becoming a serious competitor for the Li-ion cell production industry, which still has inherent development potential. It is essential to become a big player in the next generation cell production technology and not wait for miracles on a lab-scale basis, which may not survive their commercialization and cannot save lab energy densities over the path to full cells.

The "surprising conclusion" — there is no Moore and More

Surprisingly, there is no battery Moore law in the semiconductor industry. Figure 5, at first glance, suggests this because it seemed that over a period of time, doubling the electrochemical cell energy density was possible.

The essence of Fig. 5 is that within the observed period of time a continuous development of known active materials (evolution of mAh/g, less formation losses, functional coatings of active materials, optimized electrode thicknesses and their manufacturing process), separators

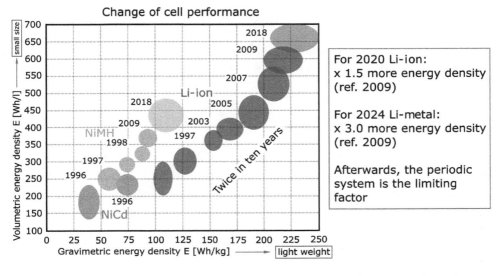

Fig. 5 Moore and More, it does not seem like it is, how does the road turn?

(minimum thickness, functional coatings), current collectors (thickness), less amounts of binders and conductive agents, low weight housings and improved electrolytes (less formation losses), have been an essential part of the Li-ion roadmap. However, the energy density can never exceed its theoretical expectation. Nonetheless, further improvements are feasible because a large variety of materials, partially not yet discovered, can be employed, but there will be an accumulation which lies at about 400 Wh/kg and 1000 Wh/l, otherwise the intercalation/insertion principle delivering cyclability, cell lifetime, safety and other important cell features becomes obsolete. So Moore and More will not continue.

Li-ion cells as long-lasting all-rounders

Alternative electrochemical systems showing up much higher energy densities on paper calculation will not save these nice values in the real world, as much as Li-ion technology is capable of performing this. The main reason is that the kinetics and the losses per cycle are much favorable if Li-ions are reversibly inserted and extracted, as in the case of a chemical conversion of sulphur or oxygen with a tremendous additional impact on very poor fast charge capability of such conversion systems. Solid-state-based systems will be more of an add-on as opposed to delivering a potential replacement option for Li-ion. This is especially true for heavy duty applications, as solid-state cells will not get rid of their inherent limitations at low temperatures. To reiterate, solid-state is not the issue, but the enhancement of energy density employing Li-metal electrodes or high voltage cathodes and a solid-state ionic conductor is only one option for this.

This function may one day also be performed by improved liquid organic electrolytes, which could be an important development for Li-ion cells. If liquid electrolytes could one day be able to sustain long-term higher voltages than 4.2 V, thus allowing one to operate metallic lithium similar to a graphite insertion electrodes and/or temperature stability up to 80°C, then energy density can be increased on a cell and system level, the latter by passive instead of active cooling. At the end of the day, only the achievable energy density on the battery level counts. Here, there is still a lot of potential to save energy density from the cell to battery level. However, Li-metal is very challenging, either in the production of ~15–20 μm thick Li-metal electrodes, which are required, or in producing roll-to-roll with them (dry rooms, sensitive to atmosphere!). In the latter, Li-ion has a distinct advantage since cell production does not require any dry room before electrolyte filling.

To satisfy the future demand for battery cells, the absolute focus has to be on safety and discovering the most quality delivering and reproducible production method. This kind of mass production is the next "high speed train", whose tickets are still available, even in Europe.

Acknowledgments

I would like to thank all my co-authors for their extraordinary work. Without their contributions, the coverage of such a wide and interesting variety of battery topics would not have been possible. Their support in the editing phase is also highly appreciated.

Deep acknowledgments go to World Scientific Publishing. Without the ideas and tremendous, continuous support of Amanda Yun and Gregory Lee, this book would not have developed into what it is now. In addition, I would like to highlight their efforts in improving the flow and readability of the text. Thanks also go to Andrea Wolf, World Scientific Publishing (Germany), for making the initial contact and accompanying me on my first steps in this book publishing journey, and to Stephan Renninger, for his support for the index and for providing valuable final corrections.

Finally, I thank my family for all their patience and encouragement during the last two years.

Stuttgart, September 19th, 2018
Kai Peter Birke

Index

48V HEV system, 187, 212
18650, 32, 40, 129, 207

absolute potential, 142
abuse tests, 53
activated carbon, 206
active carbon, 193
active material loss, 154, 156
active materials, 1
actual state, 135
adsorption model (adatom model), 191, 200
aging, 87, 97, 115, 126, 136, 141, 151, 213
    inhomogeneous, 75, 161
air cooling, 67
alkaline water electrolysis, 246
alloying reaction, 123, 124
aluminum-air, 257
amorphous carbons, 192
analytic modeling, 105
analytical model, 84
analytical transformation, 197
anode material demage, 156
anodic side reactions, 148
arbitrage business, 242
automotive application, 267
automotive standards, 52
available capacity, 146
axial, 49

balancing, 83, 101
    active, 57
    passive, 57
battery
    cost breakdown, 227

battery management system, 56, 82, 102, 128
battery module, 45
battery recycling, 227
battery thermal management, 66
Battery University, 38, 224
BET method, 207
binder material, 126
biocatalyst, 251
biogas, 251
biological methanation, 251
biomethanol plants, 255
block architecture, 45
Bockris–Devanthan–Müller model, 188
bond length, 123
bursting membrane, 37

calcination, 232
capacitive effects, 187
capacity reduction, 158
capacity variation, 84, 90, 99
carbon black, 192, 193
carbon sources, 255
cathode material damage, 158, 160
cathodic side reactions, 151
cell
    circuitry, 46
    connector, 51, 60
    housing, 31, 129
    parameter variation, 109, 116
    pressure, 125, 127, 149
    standardization, 38, 40
    supervision unit, 44, 56
    venting, 67
cell distribution, 90

cell factories, 271
cell formates, 47
cell production, 272
ceramic ion conductors, 270
charge transfer, 190
charge transfer reaction, 144
chemical precipitation, 233
chloralkali electrolysis, 246
circuit models, 195
clamped joints, 60
CO_2-extraction, 255
cobalt oxide, 223
cobalt supply, 225
coefficient of performance (COP), 74
collection schemes, 228
commercial systems, 268
complex impedance, 195
compression, 50
conductive additive, 192
conductive carbon, 125
constant-phase-element, 200
constriction resistance, 63
contact corrosion, 52
contact resistance, 51, 61, 63
    aging, 64
    roughness, 64
control strategy, 102
conversion reaction, 123, 125
cost distribution, 264
cross-sectoral energy conversion, 239, 247
current distribution, 83, 93, 97
current pulse method, 204
cyclic voltammetry, 203
cycling window, 145, 158
cylindrical cell, 31–33, 38, 40, 48, 74, 129

Debye length, 208
dendrite growth, 270
diethyl carbonate, 209
differential voltage analysis, 155
diffusion, 198
dimethyl ether, 255
disassembly of battery packs, 230
discrete optimal control, 106
discrete simulation, 111
displacement sensor, 130
double-layer capacitance, 188

dynamic optimization, 105
dynamic programming method, 107

Eddy current process, 231
electric vehicles, 205
electricity market, 242
electrochemical cell, 9
electrochemical dilatometry, 130
electrochemical impedance spectroscopy (EIS), 195
electrochemical series, 4
electrode compression, 35
electrode interface, 188
electrode loading, 144
electrode production, 193
electrode state diagram, 142, 144
electrohydraulic fragmentation, 234
electrolyte consumption, 147
electrolyte evaporation, 39
electrolytic hydrogen generation, 245
electromobility, 268
electrostatic separation, 231
end of life criterion, 213
energy density, 33, 263
energy efficiency, 103
energy loss, 104
energy storage system, 264
Epec, 38
equalization energy, 104
equalization level, 106
equivalent circuit model, 90, 91
ethylene carbonate, 149, 193, 209
excitation signal, 195

Faradaic energy, 216
Faradaic impedance, 197
film resistance, 63
Fischer–Tropsch synthesis, 253
flotation process, 231
forced convection, 54
free lattice sites, 190, 200
frequency-dependent impedance, 196
fuel cells, 243, 265

gas grid, 243
gas management systems (GMS), 258
gas venting, 37
gassing, 127, 146
gauge factor, 133

Index

Gouy–Chapman diffusive layer, 189
graphite, 11, 143, 155, 223
gravimetric energy density, 4, 243
grid stabilization, 241
GRS Batterien, 228, 229

H_2 generation on demand, 257
half-cell potential, 143, 145
half-cells, 9
heat pipe, 71
heating system, 270
Helmholtz layer, 188, 208
heterogeneous catalysis, 253
hexagonal packaging, 48
high-power cells, 206
high-temperature electrolysis, 246
host compound, 123
hybrid capacitor, 191
hybrid electric vehicles, 206
hydrogen production processes, 245
hydrometallurgical processes, 232

ideal capacitor, 196
impedance spectroscopy, 194, 195
inactivation of active material, 155
incremental capacity analysis, 155
index, 81
industrial-related development, 263
Industry 4.0, 272
infrared-communication, 58
insertion, 123
integral battery architecture, 50
intercalation, 121, 123, 143, 190
intercalation electrode, 10
intermetallic connection, 62
internal stress, 36, 134
ion exchange, 233
island network, 242

lamination, 38
laser beam welding, 60
leaching, 233
lead acid battery, 266
lead-acid, 223
Li metal anodes, 269
Li-air, 17, 266
Li-ion capacitor, 206, 215
Li-metal-ion, 17
Li-sulfur, 18

Li-sulphur, 264
LIBELLE-project, 44
$LiPF_6$, 152
Liquid cooling, 69
lithium cobalt oxide, 12
lithium iron phosphate, 12, 223, 269
lithium loss, 151, 157
lithium manganese oxide, 12, 223
lithium plating, 127, 151
lithium supply, 225
lithium titanate, 223
lithium titanium oxide, 11, 206
load cell, 134
local intermediate storage systems, 241

magnetic separation, 231
manganese, 153
mass transfer, 198
measuring the volume strain, 128
mechanical tracking, 128
mechatronic management, 128
metal compound electrode, 10
metal electrode, 10
metal hydrides, 257
methanation, 250
methanol synthesis, 253
modular architecture, 46, 102
molar capacity, 2
Monte Carlo simulation, 90, 92
multi-layer model, 202
multicomponent catalysts, 254

nail penetration, 35
negative electricity prices, 242
negative electrodes, 12
nickel cadmium, 223
nickel cobalt manganese (NCM), 143, 156, 223
nickel metal hydride, 223
nominal cell voltage, 16
non-uniform temperature distribution, 74
Nyquist plot, 195

open circuit voltage, 143
    bending, 94
    difference, 88
operation range, 82

operation temperature, 269
optical communication, 57
optimal control policy, 110
oxidation state, 124, 125

packing density, 48
parallel connection, 47, 82, 90, 97
Peltier effect, 72
periphery, 45
phase change materials (PCM), 70, 257
phase transition, 156, 203
phosphate production, 226
photo-biological hydrogen generation, 250
photochemical cycle, 249
photochemical hydrogen generation, 249
photosensizer, 249
plasma technologies, 252
plasma-based methanation, 251
plate capacitor, 189
polarization voltage, 203
polyethyleneoxide, 269
polymer electrolyte membrane electrolysis, 246
positive electrodes, 12
post lithium-ion systems, 16
pouch cell, 31–41, 74, 129
power tools, 206
Power-to-Gas, 244
Power-to-Liquid, 252
Power-to-Solid, 256
Power-to-X, 239
    conversion chains, 240
pre-processing treatments, 229
pressure map, 134
pressure tracking systems, 128
pressurized measurements, 133
primary cell, 125, 244
    chemistry, 223
printed circuit board, 57
prismatic cell, 31–36, 38, 40, 49, 129
production quality, 272
proton exchange membrane, 246
pseudocapacitance, 190, 214
purchasing cells, 271
pyrolysis, 232
pyrometallurgical processes, 232

quasi-OCV, 155

Randles circuit, 195
raw materials, 264
RC-circuit, 202
reconfigurable battery systems, 102
recycling, 223
recycling enforcement, 227
recyling technologies, 228
redox reaction, 3
resistance spot welding, 60
resistance variation, 84, 90, 99
resource access, 264
rest energy, 104

Sabatier process, 250
    reactor technologies, 250
salt depletion, 152
Sankey diagram, 103
scalable modules, 55
sealing seam, 38
second-life application, 243
selective fragmentation, 234
self passivation, 149
self-discharge, 147
self-heating, 270
serial connection, 47, 82, 102, 211
shredding, 230
sieving, 230
silicol, 258
silicon electrode, 3
simulation, 90
slag, 232
soldering, 60
solid electrolyte interface (SEI), 126, 135, 193, 208
    formation, 148
solid ion conductors, 247, 269
solid permeable interface, 151
solid-state systems, 269
solvent co-intercalation, 194
solvent extraction, 233
sorting methods, recycling, 229
specific adsorption, 189
specific capacitance, 189, 210
specific capacity, 3
specific surface, 189
square packaging, 48
stability window, 148
standard electrode potential, 4
state estimation, 82

State of Health (SoH), 113, 136
stationary battery storage, 242
stationary energy storage, 211
statistical analysis, 90
steam reforming, 248
stress-SoC curve, 137
structured-light scanning system, 133
supercapacitor, 191, 196
surface cooling, 76
surface strain, 132
synthetic fuels, 252
synthetic natural gas, 243

temperature variation, 99
temperture control, 270
terminal cooling, 75
test cells, 130
thermal management, 32, 34, 54, 97
thermal pre-treatment, 231
thermal storage, 67
thermoelectric cooler, 72
time constant, 202

titanium disulphide, 191
transition metal dissolution, 152

ultrasonic welding, 60
unpressurized measurement, 132
US Geological Survey, 224

voltage limited, 145
voltage window, 190
voltage versus Li, 142
volume strain, 121, 125
volumetric energy density, 6
    system level, 33

Warburg impedance, 196
water gas shift reactor, 253
water ingress, 39
windsifting, 230
world market, 224

ZEBRA-battery, 269
zinc-air cells, 257

Printed in the USA
CPSIA information can be obtained
at www.ICGtesting.com
JSHW060002120324
58905JS00002B/6

9 789811 215988